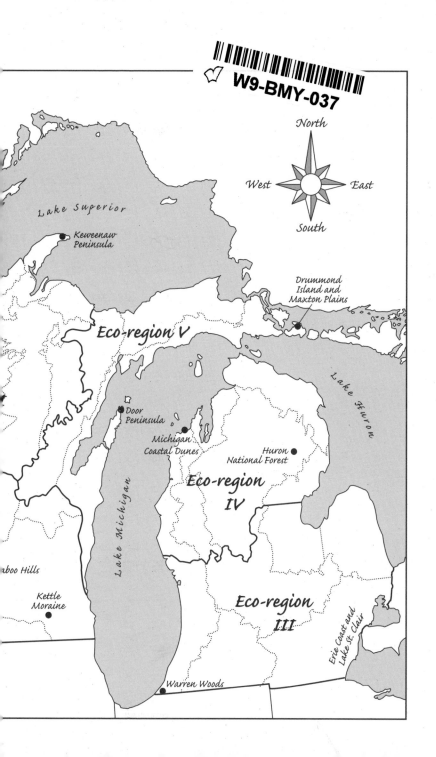

North

West · East

South

Lake Superior

Keweenaw
Peninsula

Drummond
Island and
Maxton Plains

Eco-region V

Lake Huron

Door
Peninsula

Michigan
Coastal Dunes

Huron
National Forest

Eco-region
IV

Lake Michigan

aboo Hills

Kettle
Moraine

Eco-region
III

Erie Coast and
Lake St. Clair

Warren Woods

FAR FROM TAME

The
Nature
Conservancy®

This book was made possible by support from Joanne and Philip Von Blon, and from the Minnesota Chapter of The Nature Conservancy, which the author gratefully acknowledges.

The mission of The Nature Conservancy is to preserve the plants, animals, and natural communities that represent the diversity of life on earth by protecting the lands and waters they need to survive.

The Nature Conservancy was incorporated in 1951 for scientific and educational purposes. It is the nation's leading private land conservation group, with more than 820,000 members and 1,400 preserves—the world's largest private system of nature sanctuaries. To date the Conservancy has been involved in the protection of 9 million acres of land in the United States and Canada. It has helped like-minded organizations preserve millions of acres in Latin America, the Caribbean, the Pacific, and Asia. Some Conservancy-acquired areas are sold to other conservation groups, both public and private. The Conservancy maintains an on-the-ground presence unparalleled by any other conservation organization.

The Conservancy's history of success stems from both its specific mission and its unique approach to conservation. By concentrating solely on preserving biodiversity, the Conservancy is able to set clear priorities and target its resources. Meanwhile, the Conservancy's nonconfrontational, market-oriented approach gives it the flexibility to work with a broad array of partners and to develop creative conservation strategies.

Please join The Nature Conservancy in its effort to protect the planet's natural heritage. A world diminished of its native plants and animals diminishes us all.

For more information, write The Nature Conservancy, 1815 North Lynn Street, Arlington, VA 22209, or call (703) 841-5300.

FAR FROM TAME

Reflections from the Heart of a Continent

Laurie Allmann

Illustrations by Evan Cantor

University of Minnesota Press

Minneapolis

London

The University of Minnesota Press gratefully acknowledges the generous assistance provided for the publication of this book by the Hamilton P. Traub University Press Fund.

Lyrics from Joni Mitchell, "Big Yellow Taxi," copyright 1970 Siquomb Publishing Corporation. All rights reserved. Used by permission. "Where the Sidewalk Ends," in Shel Silverstein, *Where the Sidewalk Ends: The Poems and Drawings of Shel Silverstein,* copyright 1974 HarperCollins, reprinted by permission of HarperCollins Publishers, New York, and Edite Kroll Literary Agency, Portland, Maine. "The Snow Man," from Wallace Stevens, *Collected Poems of Wallace Stevens,* copyright 1923 and renewed 1951 by Wallace Stevens. Reprinted by permission of Alfred A. Knopf, Inc., New York, and Faber and Faber Ltd., London.

Endsheet maps adapted from Dennis A. Albert, *Regional Landscape Ecosystems of Michigan, Minnesota, and Wisconsin: A Working Map and Classification.* United States Forestry Service North Central Forest Experiment Station, 1995.

Published by the University of Minnesota Press
111 Third Avenue South, Suite 290
Minneapolis, MN 55401-2520

 Printed on acid-free, recycled paper
(50% recycled / 15% postconsumer)
Text design by Rebecca Manfredini

Library of Congress Cataloging-in-Publication Data

Allmann, Laurie, 1958–
    Far from tame : reflections from the heart of a continent / Laurie Allmann.
        p.   cm.
    Includes bibliographical references.
    ISBN 0-8166-2608-1
    1. Natural history—Michigan.   2. Natural history—Minnesota.
3. Natural history—Wisconsin.   4. Landscape ecology—Michigan.
5. Landscape ecology—Minnesota.   6. Landscape ecology—Wisconsin.
I. Title.
QH105.M5A68   1996
508.77—dc20   96-14809

The University of Minnesota is an
equal-opportunity educator and employer.

For Ozzie Allmann, who pointed out the wild things
that lay beyond the window glass and said "Look,"
and for Rita Allmann, whom even the hummingbirds
recognize as a flower in disguise.

CONTENTS

Eco-region III

Eco-region IV

Eco-region V

Eco-region VI

Eco-region VII

Eco-region VIII

PREFACE

Envision the region encompassed by the states of Minnesota, Wisconsin, and Michigan. See the borderlines between the states, the delineation of counties, the checkerboard of townships. Imagine that the system is a game you invented, much like the games of childhood in which boundaries are made with chalk, sticks, abandoned shoes.

But in this game, it is an adult who spells out the rules: "Let's say this is Minnesota. We'll make this river the boundary between Minnesota and Wisconsin. And we'll say that all this land here and here, between these lakes, we'll call Michigan."

At first, you understand that the game is just a game—after all, you invented it. As time goes by, it begins to seem real. Real as rock. Real as the beating of your heart. You start to believe, when you cross one of your imaginary borders, that you are actually leaving one place and entering another.

All the while, as you are playing your game, the land is playing a different game—its own game. Its boundaries do not match yours. They are more gradual than precise, and they change over time. The land's game is played in a world that includes colors beyond your range of vision and sounds that your ears are not able to hear. It is governed by such rules as prevailing wind, the angle of the sun, the movement of water, lightning fires. You sense that you could watch the land's game forever and never know all its rules.

You look around you and see how many and varied are the players in the land's game: birds whose course takes them a thousand miles away and back to nest on the same spit of shoreline as the year before; the different kinds of prairie plants arranged on the land according to the direction faced by a slope or the character of the soil; moths following scent trails across the night to find their mates.

Although you have lived your whole life according to the rules of your own game, you cannot help but be impressed by the grace of the land's players. You admire their perseverance in continuing their game in the midst of yours. It strikes you then that it may be the only game they know. It strikes you that it is not a game.

ACKNOWLEDGMENTS

It is because of the goodwill and generosity of many people that I have been able to write this book. I am indebted to those whose names appear in the bibliography: people whose research and writings give us all a better view of the natural world; people who spent time with me in the field teaching me to see these landscapes in ways that I had never seen them before; people who enthused with me over such things as cartwheeling continents and Midwestern whales; people with whom I corresponded or spent minutes and even hours on the phone, helping me to chase after those mercurial and elusive creatures known as facts. Of the many who lent their time, expertise, support, and friendship in these ways, I wish to especially thank Lee Frelich, Robert Dana, Hannah Dunevitz, Pat Collins, Steve Wilson, Kent Gilges, Karen Noyce, Chel Anderson, and Pat Hudson.

I also extend my sincere thanks to the following people:

Denny Albert, for the inspiration that his work provided to

this book, and for all the kind and expert counsel he offered to "that writer from Minnesota."

Julie Muehlberg, Brighid Holman, Nelson T. French, Greta Hesse Gauthier, Sara Meyer, and all the past and present staff of the Minnesota Chapter of The Nature Conservancy whose efforts contributed to this project; especially Kim Chapman, who was a friend and frequent source of information, and whose passion for and knowledge of these landscapes greatly enhanced my own understanding.

To Kim Wright of the Wisconsin Chapter of The Nature Conservancy, and Dave Ewert of the Michigan Chapter, who provided encouragement, site recommendations, good contacts, and a wealth of information regarding natural areas in their respective states.

To Joanne and Phil Von Blon, whom I admire for their steady commitment to the preservation of wild lands, and whose support allowed me to devote to this book the time it required.

To Evan Cantor, whose artwork I am honored to have grace this book.

To Barbara Coffin, Todd Orjala, Becky Manfredini, Neelum Chaudhry, Mary Byers, and all those at the University of Minnesota Press whose efforts at various points along the way helped bring this project from an idea to something I can hold in my hands; and to Kathryn Grimes and the marketing staff who are working to see that I'm not the only one who does.

To my sister, Mary, and friend Susan, for help with proofreading.

To Liz Hannon, who first found room for my essays on the radio and who, bless her big heart, waits for me when I'm lost.

To my family and dear friends, especially Chris A. Baird, and with appreciation and love to John, who makes home home, and who offered incentive to come up from the basement now and then while this book was being written.

INTRODUCTION

From where do the winds come, my daughter?
How high goes the sun, my boy?
Surely these are things you know.

When I first saw an early draft of the map by Michigan ecologist Denny Albert, I looked at it for a long, long time. The map depicted the region where I had been born and lived all my life. Yet the state lines that I was accustomed to using to get my bearings were so faint as to be invisible. Superimposed over the faint outlines of Minnesota, Wisconsin, and Michigan were new boundaries enclosing states of a different kind: natural states.

With my fingertip, I traced the intriguing lines that divided the area into eight of these natural states, or *eco-regions*. Each of the eight was further divided into smaller sections of odd sizes and shapes, with names that suggested the range of elements that had been considered in the map's design: the Driftless Area,

Cassopolis Ice-Contact Ridges, Niagaran Escarpment and Lake Plain, Coteau des Prairies, Rock River Hill Country, Big Woods.

It would seem to be a paradox, but what the map represents most with all its divisions is unity. The eco-regions define large areas of relatively uniform climate and associated major natural vegetation types. The smaller sections within them define more localized landscapes whose elements are integrated enough, and share enough common forces, that they are thought to function somewhat as a whole. It is a way of understanding the now-scattered remnants of wild country in light of their native culture, a culture that has been educated by time and refined through generations.

This center of the North American continent is far from tame. The 199,000-square-mile landmass encompassed by Minnesota, Wisconsin, and Michigan—an area that could more than cover the country of Spain—has a history that includes mountains rivaling the Alps, lava flows pouring from a tear in the earth's crust, intrusions of ocean, and waves of glacial ice advancing and retreating over a span of more than two billion years.

The mountains are much diminished now, worn down to their wide bases like the teeth of old deer. Lava flows have become basalt and ocean-bed sediments have cemented to become limestone and sandstone. The glaciers left a progeny of kettle lakes and channels torn by meltwater rivers, of gravel hills and drift-covered plains from which soils and plants had to begin anew. In the wake of the retreating ice, five newborn freshwater seas—the Great Lakes—were left rocking in their basins.

Ancient forces blend with those of today to continue the "work in progress" that is every landscape. If the present seems sedate in comparison to the past it is because what we witness during our brief life spans will take on full meaning only later, when set in context with what is yet to come. Ours is the limited view available from within. But the landscapes of today are as much an arena for events, both dramatic and subtle, as they ever were.

The region encompassed by Albert's map is shaped, first and foremost, by climate. Marked seasonal change is brought about by its position halfway between the equator and the North Pole, in concert with the tilt of the planet's axis. Summers are warm. The sun's rays strike the Northern Hemisphere most directly of all the year, and days are long. Winters, the sun spends most of its time below the swollen belly of the earth, rising for as little as eight and a half hours a day to cast a weak, oblique light. Much of even this meager energy is not absorbed but rather reflected back to the atmosphere by the high albedo of a landscape covered in snow and ice. Temperatures drop low enough to turn diesel fuel to jelly and whiskey to slush.

Three major air masses travel across the continent to rule the weather. The Maritime Tropical air mass originates in the Gulf of Mexico. The Continental Polar comes from the Arctic, and the Maritime Polar comes out of the Pacific Northwest.

Contributing to their movements is a high-speed river of wind flowing aloft at five to twelve miles above the surface of the ground. As much as one hundred miles wide and two miles deep, this *jet stream* moves independent of surface winds and courses generally eastward through the midlatitudes. The position and strength of the "jet" vary with the seasonal shifts in the position of the sun relative to the surface of the Northern Hemisphere. In summer it is at its weakest and most northerly, and takes a more sedate and unwavering route around the globe. In winter it gathers strength and undulates like ribbon candy between the latitudes, often spawning low-pressure systems that bring storms.

The meanderings of the jet are important to the region because they correspond to migrations of the air masses and the relative amounts of moisture they bring. When this "wind river" takes a straight-line course out of the west, the region is dominated by the Maritime Polar, which loses most of the ocean moisture it carries as it crosses the Rockies and tends to arrive dry and mild. When the jet loops up into Canada it serves as a barrier to cold air in the north, and the warm, moist Maritime Tropical is able to sweep unimpeded up from the Gulf to cover the Great Lakes states. When instead the jet dips south, it pulls the cold and usually dry Continental Polar behind it like a drawn shade.

Adding to the climate mix are the Great Lakes. Minnesota, Wisconsin, and Michigan have a combined total of more than four thousand miles of shoreline on Lakes Superior, Michigan, Huron, and Erie. These inland freshwater seas are slow to warm in spring. They are stable under the high pressure of summer, radiate stored heat late into autumn, and are prone to winter storms when their rising moisture swells the clouds and precipitates as snow on their lee shores. In general, the presence of the lakes serves to moderate the climate in their vicinity. They extend the near-shore growing season and keep temperatures less extreme than would otherwise be found on lands at their latitude in the continental interior.

Temperature, day length, the amount of moisture carried on the winds: these are the broad parameters of climate. The way in which they play out on the land is further tempered by the character of bedrock, soil, and topography. A deep, stream-cut gorge can escape the sun's rays. A sandy soil can fail to hold the rain. From these and a kaleidoscope of other conditions has come the pattern of landscapes and natural communities that Albert sought to relate with his map. It is a pattern unique in the world — unique because the region has a history all its own and because that past moves inside the present as it does nowhere else.

A Karner blue butterfly drawing nectar from a blooming

lupine on a Mississippi River bluff savanna is being fed by the whole sequence of events that made the lupine possible—everything from the height of the sun in the sky, to patchy fire and historic grazing by bison, to the droughty soil that the river left when it was raging high with glacial meltwaters. The butterfly is the physical *expression* of these relationships.

If our hope, then, is to live among Karner blues, or among tundra swans, forests, prairies, or coastal wetlands, our first step will be to expand our ideas of these entities as best we can, to include all that they are, all that they need.

And so I take Albert's map in hand. In the space of a year I go to see again, for the first time, the world I thought I knew. I go to places within each eco-region where the wild things still gather and shift over the land like prayers set to song. I go with both mind and heart, for that is how we all go. If I have a purpose I could name, it is not to lay down a trail going in, but rather a fresh one coming out.

A Note to Readers

My primary source for information on natural communities and border justifications in the thumbnail sketches of the eco-regions that begin each chapter has been *Regional Landscape Ecosystems of Michigan, Minnesota, and Wisconsin: A Working Map and Classification,* authored by Dennis Albert and published by the North Central Forest Experiment Station, St. Paul, Minnesota. For full citations of the many contributors and sources used by Albert, please refer to the original document. Any errors in interpretation are my own.

Where I have departed from Albert's classification system for the purposes of this book, you will find it so noted. Most significantly, I have chosen to use an early draft of his map, in which the states are divided into eight, as opposed to his final eleven, eco-regions. For reference in your reading, you'll find this adapted version of Albert's map, depicted in varying levels of detail, inside the front and back covers.

The sites I focus on in this collection of essays represent a sampling of natural areas that remain in these eco-regions. While some of the sites are of a size and nature that can accommodate limited public use, others are small and fragile, or may be especially vulnerable to harm during certain seasons. Before venturing into any natural area, please seek and respect the advice of the managing agencies that work to preserve and protect these wild lands in the region. (See "A Selected List of Sites" in the rear of the book.)

ECO-REGION
I

Western Minnesota Grasslands: Tallgrass Prairie, Glacial River and Lake Features, Mountain Rainshadows

Lay of the Land

Eco-region I is characterized by flatlands and gently rolling terrain originating either as till deposited by steadily retreating glacial ice (ground moraine) or as the abandoned beds of glacial-era lakes and rivers. Isolated areas of relief are offered by the shoreline ridges left by Glacial Lake Agassiz in the north; by the broad channel carved by Glacial River Warren in the central section; and, in the southwest, by both the Bemis end moraine (deposited at the edge of the Des Moines lobe of the most recent glaciation) and the bedrock rise of the Coteau des Prairies.

Native Communities

Along with contiguous portions of the Dakotas, much of this region was historically the land of the tallgrass prairie, the easterly expression of a midcontinent band of grasslands. Few remnants exist of the true tallgrass prairie, whose grasses may attain heights of seven feet. Represented best in the grassland fragments that remain are the driest and wettest extremes: dry gravel prairies more akin to the shortgrass prairies of the west, and lowland wet prairies. Narrow floodplain forests border streams and riverways.

Animations

• Minnesota's tallgrass prairie province lies at the fading edge of a rainshadow cast by the Rocky Mountains. The mountains, by forcing air masses driven by the prevailing westerlies to rise into higher altitudes, cause their moisture to condense as rain that falls on the Pacific side of the range. Midcontinent lands to their lee are left drier and more suited to drought-adapted grasslands than to forests.

• Relatively featureless terrain, a paucity of lakes and large rivers, low winter precipitation, and drying winds combine to make this region prone to spring fires. Fires prevent succession to trees, and enrich the grassland by releasing nutrients from the remains of the previous season's growth.

• Tallgrass prairies have been called the *blacksoil* prairies. The color of the soil is an indicator of its high organic content, built slowly over thousands of years under native grassland cover. Soils that developed under these prairies are among the richest in the world.

• Dry granite outcrops in the Minnesota River Valley below Ortonville provide the right conditions for the growth of *Coryphantha vivipara,* a cactus.

• The prairie-forest border is dynamic. In a period of relatively warmer and drier conditions known as the Hypsithermal, which occurred roughly between five thousand and seven thousand years ago, continuous tallgrass prairie is considered to have extended eastward across the midsection of present-day Minnesota and beyond the St. Croix River to the vicinity of what is now Hudson, Wisconsin.

Border Justifications

This tallgrass prairie province receives greater annual precipitation than shortgrass and mixed-grass prairies to the west. Its border to the east with Eco-region II follows in part a line of increased relief associated with glacial moraines. The firebreak of these moraines, as well as a decrease in spring fires theoretically related to more winter precipitation in Eco-region II, is associated with the transition eastward from grasslands toward communities in which trees comprise a greater component.

Note

Albert divides Eco-region I in two: a northern and a southern grassland, differentiated primarily by climate and growing season shifts related to latitude.

Bluestem Prairie

April

There is nothing left to do but wait.

A train passes through the darkness outside of the blind. There is no whistle because there is no one to warn. Just the metal rhythm of wheel on track. I can't see it, but I hear the locomotive slow down as it makes the climb out of the plain of the glacial lake bed and crosses over the ancient beach ridges where the waves of Lake Agassiz once licked the shore.

This lake, when at its largest, was greater than all the Great Lakes combined. It spread over the land that would later become northwestern Minnesota, into South and North Dakota, and the Canadian provinces of Manitoba, Ontario, and Saskatchewan. Seven hundred miles of latitude. It left behind the clay-silt plain that is now the Red River Valley. And it left rows of gravel, called *strand lines* or *beach ridges,* that mark what were once the shorelines of the lake at various periods of its existence.

I remember I laughed when I first saw the ridges. "Can you see the ridges from here?" I asked. I was standing on one.

There are four main strand lines that one would cross in going, say, from Glyndon to Hawley in Clay County, Minnesota: first the Campbell, then the Tintah, the Norcross, and finally the Herman. The lowest, the Campbell, is represented in the local relief by a forty-foot rise in elevation occurring over a distance of about a mile. Five miles to the east is the Herman, the highest strand line, which gains a little over a hundred feet in elevation above the Campbell. For a person from hill country they could easily go unnoticed as another degree of flatness in a flat landscape.

Yet ask a farmer about the beach ridges, ask the native prairie, or ask a train. For a farmer these wrinkles in the land mean dry, gravel soils, dramatically poor by comparison to the lake-bed plains, and not worth the trouble. For the native prairie, long plowed under in the plains, the unfarmed ridges have meant a tenuous grip on survival, while the gravel represents an ever-present threat of mining. For a train passing in these waning hours of an April night, they mean a hesitation, an extra shot of diesel smoke into the sky, and a break in the rhythm of the wheels.

The temperature is below freezing. Since I can't jump, yell, swing my arms, I sit perfectly still and try to preserve the layer of warm air between my skin and the wool of my clothes. The wind pouring through the glassless window is cold enough to make tears stream from my eyes. I find Venus above the horizon in the east, where the sky has begun to bruise a deep midnight blue, and I take a little warmth from the sight, even though there is none there to take.

The first call comes in this time before dawn. Startling. The burst of sound comes from right outside the blind: a hollow hum that I feel as much as hear. Three minor-keyed notes shudder through the morning air like the bass tones in the timbers of a dance hall floor. It is lower than I thought it would be, and louder.

Minutes proceed in silence. These are the minutes I knew as a child, unhurried and singular, going by slowly enough that I can crawl inside each one and occupy its ripening space. I focus on the black before me, pupils dilated wide to gather the faint light. A sheet of mercury appears. It becomes a pond. The sky in the east brightens to a dull sherbet orange. The landscape of a spring prairie is slowly revealed: last year's growth of winter-worn grasses faded to the color of an old yellow lab's fur. In front of the blind are patches of bare ground and shorter grass. I pull a chill breath through my nostrils, considering that if it were only two hundred years earlier, the air would carry the rank scent of bison, elk, even grizzlies.

As if in response to some signal, the stillness shatters again, this time into a ragged chorus of voices from the clearing. The hollow humming, and now a chattering akin to that of monkeys.

I lean forward and see, a few arm's lengths away from the blind, a male greater prairie chicken, or stated more properly, a pinnated grouse. In Latin it is *Tympanuchus cupido*—loosely translated, drum of desire. Even in the dim light he is resplendent. Two crescent moons of yellow curve like eyebrows above his eyes. His chest is barred with tan and brown. Maybe two pounds all told, he is a foot and a half from the tip of his beak to the end of his short, rounded tail. Another time, he would have been one of millions in the state. Now, at the time of this writing, estimated populations based on spring surveys put him at one in fewer than five thousand.

With his feet, he pounds a short drum roll on the bare ground. Then he raises his tail, bobs his head, and long, dark feathers that had lain along the back of his neck flip up like a pair of horns. At the same instant, two vivid orange-gold balloons of skin swell at either side of his throat. These are the tympani, the resonating chambers for the wild sound that can carry as much as two miles across the prairie.

Scattered around him, in an area about a hundred feet across, I count twelve more males doing the same. This is the stage of the greater prairie chickens' courtship ritual. The booming ground. The *lek*. It is at places such as this throughout the range of the prairie chicken that groups of cocks fight their bloodless battles for status. I see pairs of them face off, defending some invisible boundaries they have made between themselves. Strutting and booming and displaying to see which of them will get the most central, the choicest piece of bare ground; which will be the two or three who will mate with any and all hens that are drawn to the lek.

I watch a lone cock a distance from the rest, doing an understated shuffle and call away from the fray of the other males. He seems to be dancing for himself. For whatever reason—inexperience, or perhaps an instinct that he can't compete with the larger males around him—he is barely engaged in the ritual. I can't help but grin and feel some empathy for him. There doesn't seem to be any great danger that his bloodline will be entering the gene pool anytime soon.

It is ten minutes past six when the sun finally peers over the horizon. The prairie chickens have settled in earnest into their work. A marbled godwit swoops in to land on the periphery of the lek. It pecks with its slender upcurved beak at a puddle and, finding it frozen, tucks up a leg under its belly feathers to rest. A harrier cruises over the lek to peruse the menu, and moves on.

Then, without warning, all proverbial hell breaks loose.

Pandemonium. The prairie chickens leap into the air, fluttering their wings and yelling out a new whooping call: fast, loud, excited. It is as though a bomb has gone off in the lek. And then I see her. A hen—small, milk chocolate brown—strolls into the clearing. She pecks occasionally at the ground, somehow managing to appear oblivious to the males. Meanwhile, they are doing absolutely everything they can to gain her attention. She makes her way toward the center of the lek, where a large male comes to meet her. He chases off any others that approach. In a few seconds he's mounted her. They separate, and with the prairie chicken's trademark flutter-and-glide, flutter-and-glide, she flies off.

She won't go far. Her ground nest will be within a mile of the lek. Four days from now she will begin to lay her fifteen hopes. One each day. As she does, the cool-season grasses of the prairie — the porcupine grass, needle and thread, junegrass — will be greening and growing taller to better conceal her nest. Tundra swans and sandhill cranes will pass overhead on their migration to the north.

When all her eggs have been laid she will begin to incubate, keeping them warm against her belly for twenty-five days with only a few half-hour breaks each day. The prairie will change around her as she waits.

She will witness the flush of the warm-season grasses in the middle of May. Big and little bluestem. Indian grass. Puccoons and golden alexanders will come into bloom. Insects will become more plentiful. By the time her young finally hatch, if they have eluded skunks and survived cold spring rains, it will be late May. Cool-season grasses will already be in flower. The hatchlings' emergence into the world will coincide with the quiet drop of birthing fawns into the prairie grass, with the first venture of fox kits out of their dens, with the brief but lovely bloom of the white ladyslipper. The young of the Richardson's ground squirrels will still be fattening in the burrows of their colonies on the beach ridges.

In the long, hot days of summer, when families of migrant workers are bent over their hoes in the sugar beet rows of the adjacent farm fields, the hen and her young will be searching for insects and seeds. They will favor areas with conditions similar to those following a fire, which offer spaces at ground level to better run and hunt. The growing young chicks will still huddle under the wings of the hen at night and whenever the elements threaten — the same set of wings providing warmth in the cold, dryness in rain, shade in heat.

By August the chicks will have mostly replaced the down of their youth with a set of juvenile feathers, fueled by the protein in the insects they have eaten. Sunflowers and the purple spikes of blazing star will sway over their heads, and the eggs of regal fritillary butterflies will appear on the violets. Velveted antlers of whitetail bucks will glide through a sea of bluestems five and six feet deep. The smoke of Manitoba fires will confuse the sky.

One day in mid-August the swallows will gather and go, then the monarchs, and, in September, the hawks. The prairie chickens will stay on through winter. When April comes again they will come back to this place to repeat the ritual. Half of those that were here the year before will have died. Those that remain will return to a booming ground to replenish what they have lost, what we have lost.

I look at the hands of my watch. It is a quarter to eight. There are long stretches now when the males are quiet. The sky is gentian blue and cloudless. An ordinary morning. For a prairie.

Hole in the Mountain Prairie

September

It appeared reather the rappid flight of birds.
(Description given by Meriwether Lewis of running pronghorns seen
upon crossing into South Dakota along the
Missouri River, 1804)

In the distance, a line of windmills turns with the power that ancient mystics knew as "God's wings." To the south are quarries where Native Americans have come for centuries to mine the layer of red catlinite stone for their ceremonial pipes. Catlinite takes its name from George Catlin, the early American painter whose palette and heart were once filled with the same colors that surround me now: the yellow and violet of compass plants, asters, blazing stars; the green, burgundy, and bronze of porcupine grass, side-oats grama, little bluestem.

Rivulets of wind flow through the prairie grasses, roughing in order to make smooth, slicing for something to mend. Nearly every day is a windy day atop the Coteau des Prairies, the prairie highlands that extend from eastern South Dakota across this corner of southwestern Minnesota and into extreme northwestern Iowa. By the time they arrive at the Coteau's rising mass of bedrock buried in glacial till, the prevailing westerlies have had

250 miles of near featureless Dakota plains within which to build speed and strength.

In another era the winds would have brought fire, pushing it eastward under a rain of cinders, sometimes even setting a bison aflame to run and roll and bawl. A single herd would stretch for more than twenty miles. When the weather was right, you could know that they were over the horizon by the great cloud of their breath. They were nomads accustomed to coming and going with the greening of the grasses that followed the fires. When they left and did not return, it was for their humps and for their tongues and for the target practice they provided through the windows of passing trains that never looked back.

Now it can seem that the only place left to look is back. The prairie that was Willa Cather's exquisite "uninterrupted mournfulness" has been interrupted. Where there was one prairie, there are now scattered pieces, all called prairies, as if the shards of a broken vase could each be called a vase.

Of the eighteen million acres of prairie that existed in Minnesota in the early 1800s, roughly 150,000 acres, less than 1 percent, are still present. Only one-third of what remains receives any level of protection, concentrated in preserves that follow this western edge of the state. Best represented among the remnants are what were once the least abundant prairie community types, those with a composition of plants associated with the wettest and driest extremes of soil moisture. Least represented in the preserves is a community whose moisture regime is intermediate between the extremes: the tallgrass—also called *mesic*—prairie, which was once the dominant prairie type of the region. Tallgrass prairie soil grows good corn.

I am but one generation removed from a farm. It is a heritage of which I am proud. Yet, as I look at the palms of my hands, I will always see with regret the trace of prairie earth caught in the lifelines. I know from whence it came.

Held up against the past, the native grassland preserves on the Coteau des Prairies are so small as to be nonexistent, inconsequential as veins of northern Minnesota gold. But we do not live in the past. Neither, clearly, do these swallows that I watch making their generous loops in the late summer sky, loops that remain for a few seconds, transparent over blue, suspended in the birds' wake. Neither in the past is the badger or coyote whose claws dug the gaping hole that takes my leg to the knee as I traverse this steep hillside, nor the Mexico-bound monarchs I see clustered on the joe-pye in the valley below.

So we begin the only way we are able. We begin not with what was, but with what is. And like the wind in the grasses, sometimes the next pass we make is one that mends.

I am on the Coteau's eastern flank. Lincoln County, Minnesota. The bedrock wedge of these highlands was once an edge against which a glacier split into two lobes. Now, laden with till and a dusting of windblown loess that brings its elevation in places to two thousand feet, it instead divides watersheds. One flank feeds into the Missouri River; the other flank, on which I stand, into the Mississippi. Before me is a valley half a mile wide, incised into the Coteau by an overflow of ponded glacial meltwater that has long since drained away. It is a place known to people of the Dakota tribe as *He hdo-ka*. Hole in the Mountain.

On the slopes is glacial till hill prairie; too dry to support tallgrass, it is akin in character to more western prairies. The lowlands are lined with wet prairie and cattail marsh. Jackrabbits sprint these hills, long legged and loose limbed. Plains pocket gophers work the soil. Upland sandpipers nest here in summer months, their wings fluttering over the heads of intruders, their calls as piercing as a hawk's cry.

A wood nymph butterfly jounces toward me on the air currents, at the last second altering its course in a rustle of wings to swerve away. If there is a calendar here, its pages are made of

butterflies. Throughout the growing season a procession of different species emerge from their cocoons in synchrony with the blooming of certain flowers upon whose nectar they feed. Spring was the Melissa blues on the locoweed. High summer was Dakota skippers and Poweshiek skipperlings on purple coneflowers. Now, in early September, the final flight brings the pawnee skippers to draw nectar from the purple spires of blazing stars.

In their own way, even these diminutive citizens of the prairies need vast expanses of native grasslands. When an isolated population in a fragment of prairie falls prey to fire, to disease in a wet spring, or to a sudden hailstorm, it can mean the loss of the species at that site unless there is a surviving population nearby from which individuals might come to recolonize the area. While the delicate-winged skippers may travel nearly a mile over inhospitable landscapes to reach an island of native grassland habitat, a greater distance may prohibit such a trek. Without large enough preserves, ideally with corridors between them, these lovely creatures will be the next to follow the bison into obscurity.

I descend the slope, then turn to face it so that its grasses frame the bottom of the sky. From this vantage there could be no world but this one, with its single white riff of clouds so stationary as to be tacked in place, feigning a permanence not in its nature.

I have come to a prairie only to find that it is impossible to go to a prairie. For a prairie is not a place, not a set of elements, but a set of relationships. It is in the moment when a butterfly catches sight of the ultraviolet pathway that it can follow down a flower petal to nectar. It is in the fine nitrogen balance that favors native grasses over exotics. It is in the reach of a root for rain, and in the posture of the ground squirrels, the "picket pins" sitting bolt upright with eyes perched high on their heads that they might peer over the grasses. It is what fills the hole of a mountain.

ECO-REGION
II

Transition Lands: Big Woods, Rivers, Savanna, and Blufflands

Lay of the Land

Eastern and western portions of Eco-region II are predominantly gently rolling terrain of glacial moraines deposited along the margins of stalled ice sheets, and till deposited by steadily retreating ice sheets (ground moraine). These are contrasted with the greater topographic relief of the central part of the region, which includes areas missed by the most recent glaciation (but affected by earlier glaciations) as well as a portion of extreme southeastern Minnesota and southwestern Wisconsin that shows no evidence of ever having experienced glacial ice. This central region is characterized by dramatic erosion features: deeply dissected ravines, exposures of older rock, complex watersheds, broad river floodplains intricately patterned by long-abandoned meanders, and terraced valleys representing successive levels of the down-cutting rivers that formed them.

Native Communities

This region is considered by some to be a transition zone from grasslands to forests, and by others as a vegetational region in its own right. Pockets of prairie are found here, but the trademark community is oak savanna: a combination of a prairie understory with dispersed, open-grown oaks, grading in places to dense oak woods. Forests of sugar maple and basswood became established where terrain offered greater protection from fires. Narrow floodplain forests line the riverways. The eastern part of the region is known for a local abundance of a groundwater-fed wetland community referred to as the calcareous fen.

Animations

• Three major rivers come together in the region as the Minnesota and the St. Croix end their run at the Mississippi. The rivers and their valleys are critical corridors of refuge for wildlife, and are seasonal migration passageways. Most intense is the autumn migration to the south, beginning in September with waterfowl and hawks, and ending in November with the exodus of bald eagles and tundra swans from Canada and northern Minnesota.

• One of the prevailing species of oak comprising the savannas and oak forests is the bur oak. Bur oaks are able to withstand regular low-intensity fire and are "plastic" in nature; that is, they are able to grow in a wide range of conditions, from extremely dry uplands to moist floodplains.

• Fragile communities known as "algific talus slopes" are found along Mississippi tributaries in association with cliffs of Paleozoic limestone and dolomite bedrock. At the base of these cliffs, cold air seeps out from ice caves and fissures in rock rubble to create a microclimate wherein even summer temperatures stay at or below forty degrees Fahrenheit. Included in the rare species that have been found at these sites are species of snails that, prior to their discovery in the algific talus slopes, had been known only from fossils dating back to the last ice age.

• A classic example of an esker, the Ripley Esker, is found within the region in Minnesota's Morrison County. Eskers are long, snaking hills made of sand and gravel that once lined the beds of meltwater rivers flowing within glaciers. Ripley Esker is more than 6 miles long, 60 feet high, and up to 250 feet across. Its north-facing slope supports a dense woodland of pin oaks, basswood, and paper birch, while its drier crest and south-facing slope support prairie and bur oak savanna. Gravel mining is the principal threat to eskers.

Border Justifications

This region is distinguished from Eco-region I to the west by having less-frequent and lower-intensity fires due to firebreaks provided by more irregular topography and rivers. Temperature differences related to latitude separate the Transition Lands from regions to the north. Lake Michigan offers a physical border to the east, on the far side of which the westerlies crossing the lake bring a more moderate climate to Michigan's Lower Peninsula.

Note

Albert divides this region into three sections, whereby the central part of the region is differentiated from lands to the east and west by its greater relief.

Nerstrand Big Woods

May

My Grandma Lucy used to tell a story about when she was a little girl. Her mother, she said, would give her a bucket of eggs to take with her on the way to school. She was supposed to stop at the general store and trade the eggs for some cheese and sausage, maybe some fruit, to have for her lunch at school. Grandma said she always traded the eggs, but she traded them for Lorna Doone cookies.

Last week I dusted off Grandma Lucy's gravestone before Memorial Day, and pulled away some of the crabgrass that had grown over the inscription: Lucy C. Stocker, February 7, 1892–February 18, 1985. She lived to the age of ninety-three, which I've always taken to be a great testimonial to cookies.

I am just outside of Nerstrand, a small village in south-central Minnesota. I can smell the manure that they spread on the newly turned farm fields. Nerstrand, translated, is said to mean "near the seacoast" or "beside the sea." It was named by a Norwegian immigrant who must have been thinking of his homeland, since the only waves here are gentle swells of land, the drift left behind by the melting of the last glacier.

I squint at the landscape of farm fields and cows and try to picture it as it was when the land surveyors came through in the mid-1800s, the way that I tried to imagine my Grandma Lucy as a young girl carrying her bucket of eggs. When the surveyors put in their section posts at every mile, they would also blaze the bark on a couple of nearby trees and inscribe the distance and angle to the post. These "bearing trees," also known as witness trees, were meant to be a backup reference just in case the post was ever moved or destroyed. In their field notes, the surveyors would record the species of the bearing trees at each section post. These notes, along with those the surveyors made in their journals as they walked their transect lines and marked the "quarter-corner" points halfway between section posts, have become important windows through which we can get at least a partial view of the composition of natural communities just prior to European settlement.

The records of the bearing trees for this region reveal the historic presence of a vast woodland of elms, sugar maples, basswood, ironwood, bitternut hickory, and butternut, with oaks around the margins. It *was* a sea—a sea of deciduous forest that lay in a long curl down the midsection of Minnesota and on into Wisconsin, Iowa, and Illinois. To the west, in lands of diminishing rainfall and more intense fires, were the prairies. To the north, the cooler climate favored a trend toward evergreen forests. But here at the prairie-forest border, in particular an eighty-by-forty-mile area in the heartland of Minnesota, was what the French explorers variously called *les Bois Francs, Bois Forts, or Bois Grands*—the Big Woods. Wolves and elk and cougar roamed under its summer canopy of leaves. It was big enough to harbor, for a while, even the hissing attack of flying serpents that were up to eight feet long and had winglike fans of skin, green underbelly and red throat. Big enough, that is, to harbor habitat for myth.

Most of the Big Woods is gone now, given way to the saw and the plow. Isolated remnants along its former range give a touch of green in the summer, some red and gold in autumn, to what has become a landscape of towns and farm fields. But on this land near Nerstrand and along the Cannon River of eastern Rice County, some larger pieces of the Big Woods have endured.

A block of more than fifteen hundred acres still exists within the boundaries of the three-thousand-acre Nerstrand Big Woods State Park. Dedicated in 1945, the park was doubled to its present size in 1992 by legislation introduced through the efforts of community groups, The Nature Conservancy, and the Minnesota Department of Natural Resources. Once a relatively small outlier of the contiguous Big Woods, these acres of forest near Nerstrand now comprise the largest remnant of *les Bois Grands* in the state.

This is the third time in as many weeks that I've come here. The first was in early May. The woods then were full of sun. Basswood leaves were still the size of nickels, and many trees were still in bud. Ferns had just begun the process of unfurling from their fiddleheads. Singing and shifting in the branches were a host of the neotropical migrants that have been called the "butterflies of the bird world" — scarlet tanagers, ovenbirds, blue-gray gnatcatchers, rose-breasted grosbeaks, and Tennessee warblers — all freshly returned from their wintering sites in Latin America. The ground was awash in the pastel colors of the ephemerals, the wildflowers that race to bloom in the brief time between the thawing of the ground in spring and the loss of the sun to the closing canopy above: bloodroots, spring beauties, meadow rue, and anemones.

Among them on a creekbank there bloomed a tiny flower that in all the world is known to grow only in the three Minnesota counties of Rice, Goodhue, and Steele. Listed by the federal government as an endangered species, the dwarf trout lily is painfully vulnerable; first, because it is a plant of limited ability to reproduce, and second, because its habit of growth by underground runners clusters the plants in such a way that a single event of disturbance could bring a major loss. For such a significant plant it makes a quiet statement in the flesh, with leaves mottled like the flanks of trout and blossoms a translucent mother-of-pearl. On a tall day, the flower stem of a single plant might achieve four inches, each inch nonetheless occupied as only a dwarf trout lily can.

When I came a second time just a week later, the leaves on the trees had become noticeably larger and more plentiful, and the forest was lit with the near-fluorescent glow of their new growth. Almost all the ephemerals had lost their petals. The dwarf trout lilies, without their flowers, were virtually invisible. Meanwhile, at the margin of the woods where the oaks grow, it was as though a blue snow had fallen. I waded through a solid bank of phlox that came to my knees. A gray tree frog surveyed the scene, its feet held tight against its body so that it formed a perfect oval little more than an inch in length. It had matched the color of its skin to the pea-pod green hue of the dogwood leaf on which it sat. If not for the glint of its gold eyes, I wouldn't have seen it at all.

Now, in late May, I arrive to find the curtain pulled closed overhead. Specks of sun are all that make it through to the forest floor. So changed is the place from what it was that I feel as though I am seeing it for the first time. This is the Big Woods of history, so deep and dark that, after an 1876 raid on a Northfield bank, Jesse James escaped the law by melting into its shadows. The wind is now something that one hears but does not feel, a soft pervasive hush through the leaves above like distant running water. Ostrich ferns rise as high as my shoulders. The air is moist and cool, full of the scent of things becoming.

What birds were moving through to the north have gone. Those that remain are intent on their own holy work—the work of being birds. A catbird tucks a piece of dried leaf into its nest. A redstart makes its erratic flights as it "hawks" for insects. When a hummingbird alights on a branch in a rare moment of stillness, I would swear I feel the earth hesitate in its turning.

Traditional measures fail to convey the real distance between this interior of the forest and its edge. For the neotropical migratory songbirds, it can represent the distance between raising and losing what is often their sole brood of a given year, and therefore, the distance between a population that can renew itself and one that cannot. These long-distance migrants comprise fully one-third of Minnesota's 234 species of regular breeding birds. Only in the depths of a large enough block of forest can they find any degree of peace from the threats that, for them, are embodied in the edge.

While predation can and does occur throughout a forest, the edge is a favorite haunt of fox, raccoons, jays, and domestic cats, all of whom readily prey upon the eggs and young of the often ground-nesting neotropical migrants. Also potentially devastating to their nesting success is the unique brand of parasitism practiced by a fellow songbird, the cowbird.

Cowbirds once roamed with bison herds to glean insects from their matted coats and from the ground disturbed by their hooves. It was not a lifestyle that allowed cowbirds to invest the time needed for traditional activities of establishing territory, nesting, and raising their young. Whether in response to this particular dilemma or some other imperative, cowbirds adopted the behavior of laying their eggs in the active nests of other species. The eggs were laid usually one or two to a nest, often after kicking out a few of the eggs of the resident pair. Sometimes the "adoptive" parents would instinctively recognize the alien eggs among their own and respond by evicting the newcomers, abandoning the nest or building a new nest on top of the old. More often, though, the cowbirds' ruse was successful and the resident pair would treat the eggs as their own.

To call the cowbird's fine adaptation "parasitism" is appropriate, for the impact of the intruders on the success of the resident brood was devastating. The young cowbirds were usually larger and more aggressive than the young of their host species. Their size advantage was enhanced by the fact that they generally hatched a day or two earlier than their nestmates. The nestling cowbirds would monopolize the food brought by the adults, and the young of the host species would often weaken and die.

Wild bison no longer roam Minnesota, but the nesting behavior of cowbirds continues unabated. They now live throughout the largely cleared landscape of North America, and have increased fourfold the number of bird species whose nests they will parasitize, particularly favoring the nests of warblers, vireos, and other small songbirds. Their history as a bird of the Great Plains is reflected in their continuing preference for the perimeter, rather than the interior, of forests. For the forest songbirds, then, the trick is to find a place to nest that allows the greatest possible distance from any edge. Seen in this light, the *configuration* of a forest's boundary may be considered as important as total area in determining its value as breeding habitat.

Breeding bird surveys conducted over the past twenty years have suggested that as many as three-quarters of the species of North America's forest-dwelling, long-distance migrants have populations in decline. In some cases, the declines are severe, with rose-breasted grosbeaks experiencing an estimated population decline of at least 40 percent and blackpoll warblers, at least 60 percent.

Accountability for the declines cannot, in good conscience, be ascribed solely to predators and the cowbird. The situation is at least in part a staging of "Guess Who We Invited to Dinner?" Only in a fragmented forest do they have more than a moderate impact on populations of other species in the community. When we maintain large blocks of forest such as this one in Nerstrand and in the wildlands of the Cannon River, then the vireos, tanagers, flycatchers, and thrushes are less likely to find undesirable surprises in their nests.

Baraboo Hills

May

The first time I was in the hills of Baraboo, I just plain didn't see them. It was a cold February morning, the kind of morning when all the lowland plants are feathered with frost. Ice crystals in the thin cirrus clouds flanked the sun with bands of color. Sun dogs.

Mine was the only car on the road. The sky so faint a blue it was almost white. Not until I was out of the hills did I see them, looking back to find the curve of snow-covered molars set in relief against the surrounding farm fields.

As I return in May, I shake my head to think that I could have passed through these hills, oblivious. Maybe a stranger needs to learn how to see a new landscape, just as blind persons whose sight is restored must learn how to interpret the perfect images that play across their retinas. Maybe the difference between looking and seeing is time.

The Baraboo Hills are what is called a monadnock: an isolated range of resistant rock rising from a plain. The word alone is cause for minor celebration. In an onomatopoetic accident of language, it reaches beyond the role of label and manages to express some of the sense of what it defines, as the sound of the word *samba* reflects the dance. To say aloud the word *monadnock* is to get a little closer to understanding the Baraboo Hills.

These hills of south-central Wisconsin are the remnant of a mountain range thought to have been uplifted, possibly by the breaking or shifting of continental plates, as many as 1.6 billion years ago. They represent the largest outcropping of a quartzite formation that extends from South Dakota to the north shore of Lake Superior. Cut a random slice from their long history and you could find them as a chain of islands in a tropical sea. Cut again and you could find them buried, rounded peaks and all, in seafloor sand.

The hills are world travelers. Roughly five hundred million years ago the part of Wisconsin that includes the Baraboo Hills lay in a shallow sea between fifteen and twenty degrees latitude south of the equator. Popularly accepted theory holds that the North American continent was at that time not only more southerly than its current location on the planet, but also turned ninety degrees clockwise so that the equator ran from what is now Texas, through Montana, to the Yukon.

By three hundred million years before the present, the continent had already pivoted counterclockwise halfway toward its present orientation. The Baraboo Hills would have had their last taste of saltwater from the retreating Paleozoic seas and would just be crossing the equator with the northwesterly migration of the continent.

Their association with the sea had overlain the Baraboo Hills with sand that, over time, became the sedimentary rocks limestone, dolomite, and sandstone. The intervening years between that distant past and the present have meant a gradual wearing down of these relatively softer rocks, exposing again the quartzite and ever deepening the canyons between the ridges. Quartzite is now revealed at the top of the hills and in sheer cliffs. The glaciers that came briefly in more recent history covered only the eastern half of the range, scouring the quartzite hilltops and leaving behind yet another veneer of sediment for wind, water, and time to whittle away.

Through it all, the quartzite has persisted. Quartzite is a metamorphic rock made of sandstone fused by pressure. It is among the hardest natural rock materials known, exceeded in hardness only by diamonds, rubies, and a very few gem minerals. The stone is lovely. A pale rosy pink. It has been quarried for gravel roads, and for ballast beneath railroad tracks because it can handle the battering of trains. Left alone in these hills of its origin, it serves an even greater purpose — the support of as many as twenty-seven natural communities with all their attendant wildlife, as well as the largest unbroken expanse of upland forest remaining in southern Wisconsin.

I had seen the statistics before I came: elevations of up to seven hundred feet above the nearby Wisconsin River; a loop of hills whose perimeter runs twenty-seven miles from east to west and fourteen miles from north to south; a wild jigsaw puzzle of ownership in Nature Conservancy preserves, Department of Natural Resources Scientific and Natural Areas, a state park, and property owned by universities, organizations, and private landowners. On a map, the ring of high country looks like a mouth with parted lips: the lower lip, or southern range, fuller than the upper. In the valley between them rests the town of Baraboo.

On the day I arrive, Baraboo's resident circus opens for the season. On the day I leave, the moon covers the midday sun in a near-total eclipse. Parenthesized between these two auspicious occasions, I wander the glens, glades, hollows, and draws of the Baraboo Hills, and find something of which I had lost sight for a while. I find hope.

I find it shaking through the body of an ovenbird singing a percussive "*k-check, k-check, k-check.*" I find it teetering in the rock-walk of a Louisiana waterthrush, who would not be here if the streams were not pure and rich with insects. I find it flying silent and heavy-bodied through the trees on the wings of a wild turkey, and lying on the ground with the heads of ginger blossoms in a rich bottomland forest. Hope illuminates the Baraboo waterways with a yield-sign yellow trail of marsh marigolds bright enough to tempt Judy Garland. It rises with every report of a bobcat, basks on bedrock with rattlesnakes, swims with pickerel frogs, and crouches with nestling vultures in the shadows of boulder fields. It unfolds from the fist-sized buds of shagbark hickories. I feel it build in my own accumulation of footfalls that, hour after hour, do not bring me to a plowed field, a road, or even a trail.

There is room here, for life. Room for deep-forest species, and for plants and animals that depend on habitats found here that are long gone from the surrounding region. In the shelter of the Baraboo Hills, at least for now, the rare are given a last chance to be almost common. And the common have a fighting chance to stay that way.

Through a rare break in the forest canopy I watch a Cooper's hawk cross the sky like a freed arrow. In its long tail and short wings are the design of a forest hunter: long tail for maneuvering, short wings to ease the passage through narrow spaces. The hawk's path is cut by the circling soar of a turkey vulture, whose wings are crafted as sails to catch the thermal updrafts of air that sweep skyward from the bluffs. Row upon row of forested hills spread out to the horizon with a continuity made all the more wondrous given the number of hands in whose ownership these lands are held. It looks strangely more like autumn than spring, with the red hues of the flowering maples and the murky orange pink of new oak leaves, still small as squirrels' ears.

It is a different forest from the one that was here prior to two hundred years ago, when frequent fires encouraged savannas of widely scattered oaks. Although still present in the hills, these more open-grown communities have steadily been replaced by those whose species can better compete in the more mesic, or moist, conditions associated with the absence of fires. By far the most dominant forest type in the hills today is the so-called southern dry-mesic, a mostly closed-canopy community whose characteristic trees include red and white oaks, sugar and red maples, basswood, walnut, and shagbark hickory. This dominance does not necessarily equate with importance. The resilience and health of "the Hills" lie in their collective diversity of communities. Prairie, open and shaded cliff communities, sedge meadows and hardwood swamps, springs and streams all shift and blend over time, each making its contribution to the whole.

I had thought to make it to this hilltop clearing before the eclipse, but I am too late. The light pours unhindered into the glade. The eclipse must have happened while I was in the narrow gorge of Hemlock Draw. In that cool and deep darkness of white pine, yellow birch, and hemlock, I would not have noticed a dimming of the light. There would have been no warning sound I could have heard above the hiss of running water in the stream; only the shadow of the moon slipping on and off the sun. It doesn't matter. There will be another one in only fifteen years.

I climb a quartzite boulder and lie across its sun-warmed surface. If anything is immutable, it would seem to be this rock. Yet even now it continues its migration on the globe, headed west with the continent at the rate of a centimeter each year. A breeze riffles through the glade. From the thin layer of soil over bedrock come a hundred stems of shooting stars. They are still in bud, but I trust that they will open when I've gone.

·

Kettle Moraine

November

The cost of breaking up an acre of oak opening
or prairie land is from $2.00 to $2.50.
(Donald McLeod, 1846)

The space I walk is a space held open by fire. It could as well be
burning now, so bright are the prairie grasses in their senescence,
shifting in the wind just as flames would shift on the ground between the spreading bur oaks of this savanna. There are the rare
kittentails, delicate sprays of prairie dropseed, Indian grass. A
colony of prairie mound-building ants swarms across the empire
it has built, intent on a thousand private missions. Spiking the
quiet are the "*nert*" calls of nuthatches reassuring themselves that
their partners are still nearby as they hunt along the crevices of
the thick fire-resistant bark of the oaks.

Reassurance is in short supply. Temperatures of twenty-four
degrees below zero are being reported from the Rocky Mountains of Wyoming. Snow has begun to appear in the forecasts of
the Midwest. The jet stream, that weather-spawning river of air
flowing eastward around the globe, is dipping back into the
lower latitudes after a summer spent over the provinces of
Canada. It is the month of the year that averages the highest

number of shipwrecks on Lake Superior. The month of the Mad Moon. The month of November.

In the hill country of extreme southeastern Wisconsin known as the Kettle Moraine, it is that hairy kind of morning that tends to keep people inside, steaming up the windows of the Dousman Café with coffee and talk. Cold rain, wind. The last of the leaves are being torn from the trees, their branches left reaching for a sun they can no longer use. Only die-hard hunters and turkeys are out wandering the backcountry. By process of elimination, I expect that makes me a turkey.

The Kettle Moraine has no borders. There is no precise moment when one could be said to pass into or out of it. If you need to know where you are, you start at the heart and work your way out. The heart, in this instance, is a low ridge of hills: the moraine. On and around it, in forty-five thousand acres of state forest land and more than two dozen designated Nature Conservancy preserves and Scientific and Natural Areas, are bogs that bloom in summer with dragon-mouth orchids, rare oak openings where equally rare Poweshiek skipperling butterflies lay their eggs on prairie dropseed, clear hardwater lakes with undisturbed shallows where blackchin shiners and starhead topminnows still swim, and marshes where bullfrogs croak literally rather than figuratively.

With the leaves mostly down it is easier to see the contours of the land that make the region a showcase for the work of the last ice age. The moraine itself runs in a snaking and discontinuous line for 120 miles through six counties from Walworth County in the south to Manitowoc in the north. It is made of drift that accumulated in what was once a corridor between two lobes of the Wisconsin glacier. On its flanks and within its valleys are drumlins, eskers, outwash plains, kames. All, at their most basic, are simply piles of rubble brought here within the ice, left to rest now on the land like pebbles from the gizzard of a great fallen bird whose carcass has long since rotted away. It was less than two hundred years ago that a handful of men stood before such landforms as these, as I am standing now, and were struck with the first explosion of understanding about the extent of the role of glaciers in the history of the continent. Makes a person wonder what we have yet to understand.

The glaciers, of course, never really left. Nothing retreated to the north. Theirs was not so much a departure as a transformation. Ice became water. It filled the depressions it had made in the land, and overflowed to spill down the Mississippi. It became the round blue eyes of kettle lakes such as Lulu Lake. It became the flooded lowlands of Horicon Marsh. It became this dampness that I feel in the wind. There is no sense getting used to it as it is. The era in which we live is considered an interglacial—a time not after glaciers but between them.

I have never been very good at imagining the past. Earlier this morning, at a museum near Eagle, Wisconsin, I ran my hand down a woody tusk nearly as long as I am tall—the tusk of a mammoth or a mastodon that once roamed these lands. With my fingers I was trying to believe: *This was once a growing thing. It belonged to an animal that would have stood as much as thirteen feet tall at the shoulder, that looked at this world through eyes like mine, and saw tundra and glaciers. It is all real. It happened.* I do the same now, as I descend the slope of the moraine, looking down its length through the open aspect of the oaks and trying to envision the two-mile-high mountains of ice that built it. But as always, it is the present that insists.

The present is a bubbling pool at the foot of the moraine. Springs, that is. Wet gold. Artesian tea. The pool is maybe six feet across, clear and shallow. Although its waters are icy cold it perks and roils as if heated from below, fed by the pressure of underground aquifers finding relief through fractures in the dolomite of the Niagaran Escarpment that runs through this

part of Wisconsin. A single black tornado of silt swirls up through the water, clearing a bright circle at its base where a fine white sediment shines through the muck.

There is no need to test the chemistry of the water in order to know that it is rich in calcium. There would be no need to test it, even if one did not know that the groundwater is steeped in dolomite or that the powdery white sediment at the base of the pool is calcium carbonate, precipitated out of the water when the carbon dioxide gas was released like fizz from a soda as the water came to the surface.

One has only to look at the plants. All the classic faces are around the table, those uncommon dwellers of seeps and springs known collectively as the community of the calcareous fen. These are the plants notorious for gathering where they can set their feet in mineral-laden groundwaters: the long blades of the beaked spike rush looping over to root again at their tips; the dried remains of an Ohio goldenrod; a lesser fringed gentian still in flower, its purple blossom rising from the yellow mound of a tussock sedge like a cup of wine on a slender stem. Here, too, I find the green basal leaves of the grass of Parnassus. This unassuming plant with the grandiose moniker takes its name from the sacred mountain in Greece where its white-striped flowers are fabled to have bloomed at the feet of Apollo, Pan, and the Muses; and where those who drink the waters of the Castalian spring are still said to be swept with a desire to write poetry.

Curious as the next person, I bend to try a taste of the water. Surely Greece has nothing on Wisconsin. After all, the National Dairy Shrine is just a few miles away in Fort Atkinson. As I reach forward, a frog leaps out from under my feet where it had been hiding under the curled lip of moss at the edge of the pool. So it is true, then, about the spring water. I had wondered where Wisconsin frogs found the inspiration for the poetry I have heard them recite on summer nights.

For the time being, I seem to have lost interest in drinking the spring water. I circle the pool to see where the frog has gone, and spot it half-buried in the sediment, relying on camouflage. It is lucky that I am a turkey and not a heron.

Something about a bit of debris in a rivulet of water leading from the pool catches my eye. Lifting it up into the air, I find that it is a tube, no longer than my thumbnail, made of hundreds of bits of grass. Such an intricate piece of work. Each piece of grass cut perfectly to size and stacked, layer upon layer, to form the triangular tube. The inside of the tube is dark. When I lower it back into the water in the palm of my hand, a tiny head immediately pokes out of one end, accompanied by a cluster of dark, thread-fine legs. The legs push against my palm to pivot the tube until it is aligned with the flow of the current from the spring. The caddis fly larva quickly secures its homemade shelter to my skin with a silky glue. Together, we wait for whatever food the current might carry past us.

In North America there are thought to be as few as three hundred calcareous fens. Of those, thirty-eight are found in southeastern Wisconsin. The secret to being rich is knowing when you are.

By the time the sun is askew in the sky I have made my way north. I have traveled out of the southern section of the moraine with its oak savannas and prairies, past a forty-mile gap in the hills where the interstate highway runs through a break in the moraine, and on into the northern section of the Kettle Moraine where richer soils and historic absence of fire have given rise to a forest of sugar maple, basswood, beech. The rain has not stopped. The wind has picked up.

At the last of the light I reach Horicon Marsh, just outside the small Wisconsin town of Waupun. Among strangers, I watch the geese come into the marsh for the night. Flocks flying in after a day of feeding in the surrounding corn fields are joined by new flocks funneling down the flyway from Hudson Bay, all using the marsh as a staging area for their journey south on migration. A man tells me through the collar of his coat, raised against the bitter wind, that airplane surveys have estimated the refuge count at 220,000 birds. Their skeins span every depth of the sky. Their calls are deafening. Somewhere behind the clouds, a mad moon prepares to rise.

ECO-REGION III

Southernmost Michigan: Moraines, Lake Plains, and Tropical Winds

Lay of the Land

Roughly sixteen thousand years ago, higher levels of Lakes Michigan and Huron edged the Lower Peninsula on both sides with broad plains of clay and sand that now extend as much as eighty miles inland from present-day shorelines. On the western side of the peninsula, prevailing west to southwesterly winds have built sand dunes within this abandoned lake plain. The dunes are concentrated within three miles of present-day shores, but are also found inland along the historic lake margins. Outside of the lake plains, Eco-region III as a whole features mostly low to rolling topography related to the deposition of till beneath and at the margins of glacial lobes.

Native Communities

Oak savannas, oak-hickory forest, and beech–sugar maple forest are the forest communities native to the interior of the region, locally interspersed with areas of tallgrass prairie. Although farming and urbanization have eliminated many natural areas in southern lower Michigan, there still remain examples of coastal plain marshes and lakeplain prairies, both globally rare communities.

Animations

• Dominance by gulf air currents, a southerly latitude, and moderation of temperature extremes by the Great Lakes combine to offer the region a long and warm growing season of up to 180 days, and milder winters than elsewhere in the three-state area.

• Rows of grape vines and roadside signs for wine tasting at local vineyards attest to the mild climate of the western coastal Great Lakes plains. Cool winds off Lake Michigan in spring help by delaying the opening of buds beyond the time when they might otherwise be damaged by frost, and the extended growing season offered by the lake's radiant heat in fall gives the grapes the time they need to ripen.

• The southern Lower Peninsula is host to at least one bona fide salt marsh, complete with species of spikerush and bulrush commonly found along ocean coasts. Brine groundwaters come to the surface from aquifers in salt deposits that originated from evaporation of Silurian seas more than four hundred million years ago. The salt marsh, just a few acres in size, is along the bank of the Maple River. The site is no secret to the whitetail deer, whose trails spread out from the marsh like spokes from the hub of a wheel, showing the popularity of the site as a salt lick.

• Lakeplain prairies are characterized by seasonal flooding. They tend to be found not along present-day coasts but inland in association with lowlands between beach ridges from historically higher lake levels. Once extensive, their distribution in southern Michigan is now very limited. Good examples still exist at the Allegan State Game Area in Allegan County and at the Sibley Prairie Complex near Detroit in Wayne County.

Border Justifications

This region is distinguished from the regions to the west and north by its overall milder climate. Also defining the region as distinct from the northern half of the Lower Peninsula (Ecoregion IV) is its relatively lower topographic relief, finer-textured soils, and natural communities of beech-maple and oak forest, oak savanna, oak-pine savanna and scattered prairie, in contrast to the pines, hardwoods, and conifer swamps to the north.

Erie Coast and Lake St. Clair

March

Limousines for Rent & Baby Potbelly Pigs for Sale
(Hand-lettered cardboard sign nailed
to a telephone pole off the Dixie Highway)

One of the bald eagles hunkers down on the large stick nest, its head just visible above the mass of sticks. The other perches on a branch just above its mate, preening the dark feathers beneath its wings. The tree cradling the nest is rooted in a plain that was once beneath the postglacial waters of a swollen Lake Erie. Now the lakeshore is more than a mile away to the east.

≈

The final day of March. In the company of muskrats, killdeer, male red-winged blackbirds, a pair of nesting bald eagles, and countless thousands of waterfowl that filled the sky like swarms of insects, I have spent the past weeks along the southeastern coasts of Michigan. Slowly, I worked my way from the shores of Lake Erie at the state's border with Ohio up to the surreal industrial complex at Sarnia known as "Chemical Valley." I was naive enough to be surprised by a landscape in which most of what passes for wetlands has been neatly partitioned into rectangles by raised dikes. Water level, once in the hands of season and wind, rain and snow, has become a matter decided by the setting of pumps.

Until this moment I had nearly given up hope of finding more than a glimmer of the region's marsh, wet prairie, and swamp forest that had extended as much as five miles inland on historic lake plains and up riverways prior to the Swamp Act of 1850.

But before me, cupped between the shores of Ontario and Michigan, is Lake St. Clair. This youngest sibling of the Laurentian Great Lakes system took its existing form only 3,200 years ago. The runt of the litter, Lake St. Clair is roughly thirty miles across at its widest point and has a mean depth of less than ten feet. It is fed from above by the waters of Lake Huron via the short strait of the St. Clair River, and drained by the Detroit River toward Lake Erie to the south. Its color is turquoise owing to the refraction of sunlight through the calcium carbonate it carries in solution from Lake Huron's limestone bed.

Its shape bears an unsettling resemblance to the human heart.

I am on the northern perimeter, at the point where the single braid of the St. Clair River begins to unravel into separate strands before entering the lake. Between the half-dozen or so channels of water are islands and marshes built on the mostly sandy sediments that the flowing waters drop as they meet the slower currents of the lake. It is a classic bird's foot delta, like the delta of the Mississippi in the deep south of Louisiana.

While highways and highrises of Detroit pulse on the lake's southwestern shores, this largest freshwater delta in the world pulses with a different kind of energy. Although a far cry from pristine, along its waterways are still areas where plants may make their own subtle responses to their environment. Oak and ash forests claim the higher ground, their branches barren and gray on these first days of spring. Then there is a transition lake-ward to dogwood meadow. Then to the bright gold of cattail marshes and finally, in the deeper waters, to bulrushes.

In the midst of a land that could hardly be more altered is this pocket of relatively wild country—a place not only where a quarter of a million ducks can be seen resting and feeding during fall migration on a single day count, but also where rare king rails still nest in summer, their calls clacking from the marshes. A place where white flurries in November are not snow, but tundra swans; where island trees house the rookeries of black-crowned night herons and egrets, and where still there is felt the occasional belly-touch of an eastern fox snake or a massasauga rattler.

Fish that elsewhere along the coasts find themselves cut off from ancient spawning grounds by the flip-flap valves of dikes are able to make their passage unhindered up the delta's tapering channels into the shallows. More than forty species of fish use the area for spawning and nursery beds: bass and yellow perch, northern pike, sturgeon and muskies.

I watch a flock of canvasbacks fly over the lake, their bodies suspended between the blur of their wings. They circle, sun flashing off their pale upper body feathers, while the lake below them grows steadily younger.

Water entering Lake St. Clair takes only nine days to circulate in its basin before moving on down the line. That is to say, if I returned in nine days, the lake would appear just as it does now, but would be an entirely new body of water. This swift exchange has tempered the fate that the system might otherwise have known with its shallow basin and small size. The short retention time (as compared to three years for Lake Erie and nearly two hundred years for Lake Superior) helps to reduce the concentration of contaminants and avoid the overenrichment of the waters by runoff from adjacent farm fields.

It does little, however, to shelter the lake from the threat posed by a recent immigrant to its waters. Lake St. Clair, in addition to being notorious (and extremely popular) for collusion in the transport of fine Canadian whiskey to American shores during

Prohibition, also has the dubious distinction of being the site within the Laurentian Great Lakes where the first zebra mussel was discovered in 1988.

These quarter-sized mussels are thought to have made their way from Europe into Lake St. Clair by the discharge of freshwater ballast from an oceangoing vessel. The vessel would have passed through the St. Lawrence Seaway, the dredged commercial shipping channel that makes it possible for large vessels to travel from the St. Lawrence River throughout the Great Lakes. Zebra mussels are to a freshwater lake community what leafy spurge is to a tallgrass prairie: an invasive exotic with astounding reproductive capacity that has the potential to displace native species and dramatically alter the function and diversity of natural systems.

By hitchhiking on the hulls of boats, zebra mussels have moved through the chain of lakes to the Duluth Harbor of Lake Superior. In Lake Erie they can now be found in concentrations of as much as seven hundred thousand per square meter.

The effects on Lake St. Clair are not fully known. The water has grown clearer as the exploding population of mussels siphons more and more nutrients out of the water column. Since the nutrients are part of the same food chain that supports pike, bass, walleye, and bluegills, there is concern that these fish species may be adversely affected by the competition. The increased clarity of the water and associated sunlight penetration may allow more rooted plants to establish, the decay of which could bring oxygen depletion when the lake is under winter ice. And the energy structure of the lake is changing. As the zebra mussels take suspended organic material out of the water and redeposit it as waste on the lake bed, they are potentially favoring bottom-feeding species such as channel catfish and carp.

Greatest thus far has been the devastation of the lake's once diverse community of native freshwater mussels, encrusted and smothered in their beds by the zebra mussel. As recently as 1986 there were at least seventeen species of native mussels (also called unionid mollusks) known to inhabit Lake St. Clair. Today researchers are hard pressed to find a single one.

Their loss may not be significant to the gulls, bluebills, and muskrats who dine on the abundant zebra mussel as readily as they did the native species. But a loss is no less a loss just because there are others of a different breed who persevere. Ten years ago I could have stood at the edge of this turquoise lake while a native pocketbook mussel beneath the water waggled a flap of its shiny flesh in perfect mimicry of a minnow to entice the approach of its host fish—a bass or freshwater drum—which would have unwittingly swum away with the offspring of the pocketbook attached to its scales.

Now the closest I can get is to recite their names. Giant Floater. Spike. Pig-toe. Threeridge. Fat Mucket. White Heelsplitter. Pink Heelsplitter. Wavy-backed Lampmussel. Fragile Papershell. Black Sandshell. Deer-toe. Mapleleaf. Strange Floater. Round Hickory Nut. Fawnfoot. Belated additions to the cost of the St. Lawrence Seaway.

In *Jane Eyre,* Charlotte Brontë wrote that remorse is the poison of life. I do not agree. Far more dangerous than remorse is its absence. For with remorse comes accountability, and within accountability there is hope for what remains.

The eagle on the nest rises slightly and reaches under its belly, perhaps to turn an egg or check on a downy nestling, then settles back down. Its mate swoops off the branch and moves with slow but sure wingbeats toward the lake.

Warren Woods

April

Start with a low, rubbery quack. Amplify it and multiply the number of sources until it is a din that can be heard a quarter-mile away. Then place the sound deep in a forest, emanating from an ephemeral pool, a shallow temporary pond made of snowmelt that will be gone in a matter of weeks.

❦

I follow the sound through the forest until I am at the edge of a pool of water, where my presence is greeted with abrupt silence. Five minutes. Ten minutes. Thirteen minutes pass before a solitary note is offered again. It is joined by another, and another. The surface of the water breaks out in small, dark bumps.

Slowly, *slowly* I raise my binoculars. I train them on a bump only a few feet away and find myself looking into a pair of shining eyes. Wood frog eyes.

After a hundred frogs I stop counting. The water begins to swirl. Male frogs lunging after the larger females, trying to mount them before they eddy away. Usually failing. A female eludes the

49

grasp of a male. He immediately resumes calling. Even days ago, he could have been taken for dead, hibernating with a barely beating heart in some muddy bottomland. I watch him, see the air sacs on the side of his head inflate and deflate as he calls. His legs float loosely in the water.

They speak well of these woods, the frogs. To ancient Egyptians, it was a frog that served as midwife at the birth of the world. Frogs have been symbols of purification, healing, harmony, fertility. I contend that a frog is fine enough as a symbol of itself. But in these times the presence of a frog has taken on new meaning. Since the eggs of amphibians are particularly sensitive to pollution, they have become indicators of water quality. An abundance of frogs is an indicator that the waters of a place have not become too polluted to sustain them. A spring cacophony of frog song is a sign of health and hope for us all. As well as a fine symbol of frog.

This place goes by the name of Warren Woods. Tucked in the southwestern corner of Michigan, it is a remnant of the beech–sugar maple forest community that once covered thousands of acres in the Lower Peninsula. All told, there are three hundred acres of mature forest here, with some of the largest and oldest beech and sugar maples found in the state—trees that rise 125 feet in height, often with diameters greater than four feet; trees that have seen as many as 450 years come and go.

Spring arrives later here than it does at the same latitude in the state's interior. It is held at bay for about three weeks because of the cold winds coming off Lake Michigan. On this day in early April, I awoke at dawn to fog. Noon brought rain. An hour later it was snowing, and now sun has broken out and the sky is a half-hearted blue. The wood frogs fall silent when I move away from the edge of the pool. But I haven't gone far when I hear their calls start up again behind me. They must, it occurs to me, like their work.

I walk for a while through the matted drift of last year's leaves that blanket the ground between the widely spaced pillars of trees. The trunks rise uninterrupted until they are far above the forest floor, then break out into branches that meet to form a continuous arching canopy like cathedral ceilings I remember from Italy, minus the heads of saints. The buds on the trees have not yet opened.

Soon the understory will flush with wildflowers: prairie trillium, spring beauties, Dutchman's breeches, and beechdrops. By July the flowers will be gone. The summer woods will be made of shadow and dew and a thousand shades of green. October will flame with yellow beech leaves and red sugar maples. Lovely seasons, all. Still I would trade no other for this time in winter's wake, when the colors are solemn and the way is muddy, and spring is but a promise you can smell in the air. It smells of worms and damp wood, of newly thawed soil.

I come upon a stream bank and look down to see the Galien River flowing swiftly along its course. The time is right for steelhead trout to be spawning. Steelhead are not native here; they were introduced to the Great Lakes from the Pacific over a hundred years ago, in the mid-1800s. From the vantage of a fallen log extended over the river, I watch for sight of their sleek bodies.

Both male and female steelhead move up the Galien from Lake Michigan. The run begins in late fall and is spread over the months until early spring, which is the peak of spawning. The male chooses a spot in the gravel bed and swims wildly in place, fanning up silt that the current will sweep downstream. He will wait for a female to select the depression he has made as a place to lay her eggs and then will fertilize the eggs as she lays them.

The Galien is not the best of nursery streams for trout. There are predators of the hatchling trout here in the warm waters of the south that would not be found in such numbers in the colder streams of the north country: walleye and perch, smallmouth bass and channel catfish. Only a few young trout will live long enough to see Lake Michigan.

If there are fish swimming below me, I cannot see them. The surface of the water has a sheen that gives back nothing but the quavering reflection of trees along the bank and the featureless moon of my own face.

Continuing along the river's narrow floodplain shelf, I scare up what appears to be the same pair of wood ducks at every other bend. This mature forest, with its many tree cavities, offers them ideal nesting habitat. The old sycamores that lean over the river are riddled with holes. I watch the male wood duck in his breeding finery of emerald green, slicked-back crest, and chestnut breast. The female is harder to see against the river bank in her soft gray-and-brown plumage and white-ringed eyes. The two of them have a great capacity for surprise. Each time they see me they react anew with frenzied calling and a splashing rush of wings as they make their escape upstream, only to have me appear again minutes later. I head away from the river and back into the woods to give them some peace.

Some of the tree species of Warren Woods are old acquaintances; others are new to me. The dominants are beech and sugar maple, easily known. The beech has smooth, silvery gray bark and long twigs as fine as the bones in a bat's wing. The bark of the sugar maple has a yellow undertone, and twigs that arise on the branches directly opposite each other. (Although I was once told, upon asking how to identify a sugar maple, that I should simply look for a tree that looks like one.)

Among the sprinkling of trees that occur in smaller numbers are sycamores, tulip trees, and black cherries. The sycamore, with its trademark "paint-by-number" bark of blotchy gray, pale green, and cream, is limited to the banks of the Galien where it can get more sun. The trunk of the tulip tree rises true and uniform as a ship's mast. It is the largest of the trees referred to as the

eastern hardwoods, said to grow to a diameter of up to sixteen feet, though not here. A young black cherry tree is told by its smooth bark with white freckles called lenticels through which air is exchanged. As the tree ages the bark becomes darker, with scaly flakes. In late summer when the fruit is ripe, a black cherry tree is the one with nearly as many birds as leaves.

For years I had a black-and-white photograph on my desk of a child walking alone through a forest of great trees. The child's back was to the camera, her head inclined upward. I always wanted to see the expression on her face. Now I don't need to see it. I think that I am wearing it.

It is not only the size of the trees. It is the assembly of species. The stark rise of their trunks against the sky. The club moss growing up tender and new and startlingly green against the backdrop of bare spring soil and scattered leaves. It is the long line of sight, the open aspect afforded by the lack of any significant shrub layer. It is the sad rarity of stands with the integrity of this one.

And it is the knowledge that a beech–sugar maple forest is considered to be a climax forest in this region. Over time, one forest community is generally replaced by another of different species and composition. But a climax forest, barring some great disturbance that would set it back to an earlier successional stage, is the theoretical end of the line. Both maple and beech saplings have the uncommon ability to hang on, at least for a while, in dense shade. They can outcompete other species under the closed summer canopy of their parent trees. Thus, a beech and sugar maple forest has the potential to be succeeded by a beech and sugar maple forest ad infinitum. Or at least to perpetuate itself to the extent that anything can in a world that, on a broader scale, is remade continuously.

Everywhere on this forest floor I see the saplings of maples and beech. They will grow only slowly under the suppression of shade. It will take the felling of a mature tree by disease or storm, and the sun that will stream through the new opening, for these young trees to be able to make a break for the sky. One such event may not offer enough time for a sapling to attain the canopy before the gap closes over once more. Caught again in shade, the sapling must then await another chance at sun. It is a chance that may or may not come in its lifetime. That it sometimes does is evidenced by the annual rings in the large trees, which tend to show a pattern of spurts of growth, theoretically related to a succession of openings that the tree experienced on its way to the top. Curiously, there are many more maple saplings than beech in Warren Woods, while there are many more beech than maple among the large trees. Research has suggested that beech saplings may be better able than the maples to endure the often extended periods between events of a tree fall.

I get a sudden glimpse of some future April day when I and everyone I know are long gone, but this forest still thrives under the shade that some of these saplings will one day live to cast. The trees will look much like these—silvery beeches beside butter-toned maples. The air will reek of sweet decay. A dozen wood duck eggs will lie on a bed of their mother's plucked feathers in the cavity of a tall sycamore. The Galien will run with trout that no one will see. It will be a day that begins with fog, moves on to rain, snow, and, finally, sun.

And if that male wood frog manages to get a little lucky today, there ought to be some music.

ECO-REGION IV

Michigan's Lake-influenced Northern Lower Peninsula: High Plains, Frost Pockets, Dunes, and Barrens

Lay of the Land

The northern part of Michigan's Lower Peninsula is characterized by diverse topography, featuring a high central sand plateau formed of moraines and glacial outwash (till carried in meltwater), towering coastal sand dunes in the west, islands, and abandoned lake plains of clay and sand on the perimeter of the peninsula formed by historically higher levels of Lakes Michigan and Huron.

Native Communities

Native communities include northern hardwood forests (composed of beech, sugar maple, yellow birch, and eastern hemlock), jack pine barrens with northern pin oaks, white and red pine forests, dune communities and interdunal wetlands, hardwood conifer swamps, and conifer (mostly white cedar) swamps.

Animations

• In contrast with Eco-region IV's coastal, more lake-moderated climates, the interior High Plains are host to extreme and variable temperatures, where winter lows can drop to minus fifty degrees Fahrenheit. Depressions formed in outwash sand by glacial ice blocks have become frost pockets where deciduous trees cannot compete with the more cold-tolerant jack pine and dry prairie communities.

• A handful of sites along northern Lake Michigan shorelines and island beaches are among the last few places in the Great Lakes basin where the endangered piping plover still makes its nest. Their nest is simple: a scrape in the sand that the birds line with bits of driftwood, pebbles, and shells.

• Fire plays a major role in maintaining the pine barrens community. Charcoal found preserved in bogs provides clues to fires of the distant past, while standing trees that have survived low-intensity fires record more recent fire history in their rings. Both the year and the direction of fires can be read in a tree's rings. Air currents tend to split and curl around a tree's trunk, bringing the most intense heat not to the side of the tree facing the oncoming fire, but to the backside. The direction of a fire's origin is thus revealed as being *opposite* that of the charcoaled portion of the tree's rings.

• In the late 1800s, the ecologist Henry Cowles made note of what he variously called *tree cemeteries* or *ghost forests* of Michigan's dunes. He was referring to the twisted stands of cedars that are sometimes unearthed at the bases of migrating dunes, resurrected hundreds of years after they were first overtaken by the sand that caused their demise.

Border Justifications

Serving as the region's northern boundary are the Straits of Mackinac, the water passage from Lake Michigan into Lake Huron that separates Michigan's Upper and Lower Peninsulas. The region is distinguished from the southern Lower Peninsula (Eco-region III) by its more variable and overall cooler climate; by its dominance of northern rather than southern air masses; by the greater amount of lake-effect precipitation received as snow rather than as rain; by its large topographic features and higher elevation; and by the shift in plant communities associated with these conditions. The region is distinguished from landmasses to the west, across Lake Michigan, by having a more lake-moderated climate and abundant lake-effect precipitation.

Michigan Coastal Dunes

March

The further we sejourned the delightfuller the land was to us.
(Pierre Radisson)

It is five in the afternoon. The sun is still a hand's width above the line between water and sky that serves here as the western horizon. I squint at the glare but decide against putting on sunglasses. In March, squinting at the sun is something midwesterners enjoy. I sit with bent legs, arms resting on my knees. My feet, bare in the beach sand, are almost warm. It is fifty degrees. On the eastern shores of Lake Michigan, in the dunes of the Leelanau, I am quietly pushing the envelope of spring.

I scan the lake for whales. A blow spout. Maybe a flash of dorsal fin or tail. The remains of at least three whales have been found buried in the Lower Peninsula of Michigan. Ribs and vertebrae of bowhead, sperm, and fin whales. Since the first bones were discovered over a hundred years ago, scientists have speculated about their origin; about when and how the whales might have made their way inland from the saltwater seas. Favorite theories have linked the presence of the whales to a period during and following the most recent series of glaciers. Rivers were swollen with meltwater. A succession of immense lakes, predecessors to the modern-day Great Lakes, covered the land as the glaciers advanced and retreated. There was water enough, presumably, to accommodate the passage of a few vagabond whales.

Given the sites where the bones were found, the fin whale was associated with Glacial Lake Arkona, and the sperm whale, with Glacial Lake Whittlesey. Both of these glacial-era lakes are considered to have been in existence about thirteen thousand years ago. The bowhead whale was said to swim in the waters of postglacial Lake Nipissing only four thousand years ago. This is not some blurry ancient past. Four thousand years is imaginable — the collective lifespans of fifty grandmothers.

The thought of whales swimming in these inland seas is a wondrous image, one that I have turned over again and again in my mind since learning of the bones. It is surprising and yet not surprising, given that the lakes have always seemed to have a kind of sentience that the presence of whales in their ancestry might explain.

But the years of speculation took a sudden turn in the late 1980s. In a published report by C. R. Harington, an Ontario paleontologist, it was announced that tests conducted on the various bones had revealed that the whales had died between 190 and 810 years ago. Since no natural conditions existed during such recent history that could support any reasonable route these whales might have taken from the sea during their lifetimes, Harington's conclusion was that the bones must have been carried by people from the coasts into the interior. Native Americans of the Hopewell Culture, he postulates, may have brought them along trade routes from the Atlantic.

Still, out of habit, I search the surface of the lake for sign of them and am not disappointed. I am successful in looking for whales.

I let my eyes play over the lake, the stretch of beach. An undulating row of low sand dunes, maybe thirty feet high, runs roughly parallel to the shore. Inland behind the dunes is a dip down to the water table, a so-called interdunal wetland, and then an ascent to a second, far higher, set of dunes.

These sand dunes, and not whales, are the true legacy of those immense lakes of the past. It was Lake Nipissing, that last of the Great Lakes' predecessors, that set the level of many of the higher dunes in this system of coastal sand dunes along Lake Michigan's southern and eastern boundaries. Driven by prevailing westerly winds, the waves of Nipissing brought great volumes of sand to the near-shore region. When receding waters exposed the sand, the wind built of it these high dunes rising from beaches and perched on top of moraine bluffs to heights as much as 450 feet above the current level of the lake. Together with the smaller foredunes born of more recent times, they form a band of sand that reaches as much as five miles inland and nearly continuously up the western edge of the state of Michigan. So vast an expanse of sand is it that astronauts have easily seen it from space. It is the most extensive system of freshwater sand dunes in the world.

In all the Great Lakes basin, as defined by the five lakes and the area encompassed by their collective watershed, the coastal dune system is the richest in endemic species. An endemic species is one confined to, or exclusive to, an area. The dynamics of wind, wave, dune topography, climate, and changing lake level have brought about a unique breed of life in these lands that hug the shores. Plants such as the Pitcher's thistle and Houghton's goldenrod, and animal life such as the Lake Huron locust, are found only among the dunes of the Great Lakes.

It is a landscape remade continuously. The adage of mountain weather applies: "You don't like it? Wait a few minutes." The same winds that played with the sands of the Nipissing era are still pouring from the southwest and northwest, gathering strength across the broad fetch of Lake Michigan, pushing waves and changing the shape of the land and the life that clings to it. The scene around me is nothing if not wind made visible. The dunes incline in their slopes. An osprey holds stationary in the sky against a stiff headwind, wings outspread. A bent blade of dunegrass swings back and forth, its tip etching concentric circles in the sand. Even the branches of pines on the most inland ridges are grown like flags blown east.

I take up a handful of beach sand. The grains are mostly quartz crystals that shine translucent in my palm, peppered with the darker minerals of hornblende, garnet, black magnetite. Each is an emigrant of some far shore or the bank of some tributary stream that feeds the lake. These grains represent *nourishment* to the dune community, whose existence depends on the continued resupply of what the winds take away. Their edges, once sharp, have been worn round by their histories. Their simple existence is a testament to their ability to endure where other rocks have gone to dust. I scatter the grains of sand back onto the beach, then rise to walk.

The rays of sun are hitting the beach with horizontal light. It is the kind of light that gives the fleeting clarity of a funeral, casting the landscape in an exacting dichotomy of light and darkness. Every westward, windward face is illuminated. Every lee side is dropped into shadow. Half of my body is warm. The other half, chilled. The topography of the shore is accentuated. Textures emerge. I become aware of a pattern repeated in the alignment of the waves, the faint corduroy ripples in the beach sand, and the deeper swells of the dunes.

I go to the water's edge, where my feet are shocked into apathy at the wash of the first wave. The ice went out only three weeks ago. As I watch the waves roll in and break offshore, I remind myself that I am not seeing water moving forward but rather a force moving *through* water as a shiver would pass through my body. I gauge the distance between the crests of the waves at fifteen, say sixteen feet, then divide that distance in half. My result, eight feet, is the estimated depth of each wave's significant influence under water: the invisible vertical wall that moves as a shiver of force through the water below each crest I see on the surface.

At the point where a wave breaks offshore, I know that the foot of its underwater wall, eight feet down, is tripping on the rising slope of the lakebed. The sand it kicks up in the process is free to find its way to shore or be carried in longshore currents to be deposited elsewhere. The makings of dunes.

Turning my back to the lake and setting out at a right angle from the shoreline, I cross first a barren band of beach too windswept and worried by waves for plants to be able to take hold. Just beyond it is the storm beach, sometimes called the middle beach. Its inland extent is marked by a ragged line of wave-tossed debris, where I find remnants of aquatic plants, driftwood, drowned insects, fish bones, and sodden feathers. This is the cafeteria line of shorebirds and tiger beetles, and of creatures such as raccoons who come in the night to dine on the day's fresh (or not so fresh) arrivals. The storm beach is temporary home to plants only in summer, when annuals such as seaside spurge and sea rocket rush to set seed before the inevitable storm waves have the chance to tear them from the sand.

Moving still farther inland, I encounter the gently rising slope of the foredune. The location, shape, and height of these young foredunes depend on the amount of available sand, and on obstacles presented to the wind. A plant, a fence, a tree, a person with limited ambition—anything that stays put long enough to trap or slow the blowing sand can form the foundation for a dune.

It is on the lake-facing slope of the foredune, maybe twenty paces from the water, that I come upon the first perennial plants that have managed to establish their roots in the loose sand. Yellow tufts of marram grass. At first widely scattered, the tufts become increasingly dense as I proceed up the face of the foredune. These could be considered tough conditions for a plant:

little surface moisture in the soil, extremes of cold and heat, and drying winds. It is capricious ground, with a bent for travel.

Marram grass is one of the few plants up to the challenge. Its leaves can be buried again and again by drifting sand, only to respond with new growth from underground stems that can send new leaves to the surface. Where the grass succeeds in stabilizing the sand, over time, it makes a dune habitable for other plants. Little bluestem grass, sand cherry trees, oaks, and pines each follow as their varying conditions for survival are met.

The skin of vegetation has, at best, a tenuous hold. If disturbance should remove any of the foredune plants, the wind is given a notch through which to pass to the backdune. Once given access it can quickly sweep away whatever small amount of organic soil has accumulated, and carve a great hollow called a blowout. A blowout takes on a life of its own. Its crest migrates inland as sand blown from the windward side lands on the lee side, or slip face. Everything in its path is swallowed. I have seen the slip face of such a blowout, seen the sand rising around the trunks of trees as a snake would walk its jaws around a still-living mouse. Cottonwoods are among the few trees that can take it, and only for a while. They grow fast and send out new roots from their trunks when overtaken by sand. Other trees die long before the sand reaches their topmost branches.

A migrating blowout can quickly undo a thousand years of progress made by the plants that colonize a dune. Few things are harder to stop than a blowout on the move. Few things are easier to start.

From my present vantage on the crest of a foredune, I look at the seemingly stable forest of a backdune. There are traces of snow on the north-facing slopes. A brown creeper spirals up an oak tree. I hear the voices of golden crowned kinglets just returned from their southern wintering grounds. If this scene is to be obliterated, swept clear of plants and consumed by sand, I want it to be from a great storm off the lake. Maybe from the waves of a June thunderstorm with lightning that fuses the sand into daggers of glass. Or from the steady assault of the strongest winds of the year that come in January. I take a long time to descend the dune. It always takes longer when you choose where each foot will land.

At first I don't notice the gulls. They fly overhead, strangely silent, pouring in a thick stream from the ridge of the backdune and out over the lake toward the sinking sun. I wait, but there is no end to them. At least no end that I can see.

I know there is a difference.

Huron National Forest

July

There is an island of sorts in the landlocked center of Michigan's Lower Peninsula. It is an island of cold air settled over the high plains of the Huron National Forest. At night the air runs down the slopes of the Maltby moraines and settles in the old glacial channels and lake basins. People here tell of how the killing frosts can come every month of the year. Right on through summer.

For the most part, it's not the kind of thing you notice on a July day. Hot still feels hot. But it makes itself known in the temperature averages and extremes, and in a shortening of the growing season. Translate these quiet circumstances to a plant, then add the bone-dry soils of grayling sand, and it becomes a shout, a riot, a howl. In the natural community, it is enough to make the difference between forest and barrens. Enough to keep out all the oaks except the northern pin oak, and to open the door for jack pine. Enough to delay the green-up of the grasses in spring that, historically at least, meant one thing: fire—small, fast fires on a regular basis that would take out the understory and leave some mature trees alive, recording their passage as a charcoaled layer in the trees' annual rings. Every forty years or so there'd be a bigger, hotter fire that would take out all the trees and clear the way for a new forest like the one I see before me now.

It's called a "dog's hair" stand of jack pine, and it looks just like its name. The trees come up thick like this, small and plentiful, when the land is cleared and the jack pines are the first to lay claim to the exposed soil with the seeds of their fire-opened cones. The peer pressure of all their same-aged contemporaries keeps the trees small. The trees of this stand are fourteen years old, yet I could still encircle almost any given trunk with my two hands. Their heights all hover at about seven feet. They grow so close together that their branches interlace. It's not the kind of place that looks remotely inviting to walk through, which is just as well. It is better that no one walk there.

This forest has already been spoken for. Or, more accurately, it has already been sung for.

It belongs to the Kirtland's warbler. Not forever; just for ten years or so, when the jack pines are between about eight and about eighteen years of age; and within each year, only for the months of May through August, when the warblers breed. The rest of the year they are either in migration or are shifting and muttering through the scrubland of the Bahamas. Except for a few isolated sightings of migrants and apparently nonbreeding birds, that's it—the entire known scope of the species' limited range: the Bahamas, a handful of counties in Michigan where they nest, and the air in between.

The Kirtland's is among the rarest of birds on the planet. I don't say this to mean that they are intrinsically more important than the Lincoln's sparrow whose song I hear at this moment, or than the ravens I saw earlier, hunched under the weight of their morning. Rareness speaks to the urgency of the call for action on our part. But considering the reach of the universe, any life is rare enough to turn my head.

At the time of this writing there are now thought to be 1,266 Kirtland's warblers drawing breath. That's 633 males counted while singing, and an inferred mate for each. At best it is an estimate, since some males have more than one mate and others are bachelors. Since people first began paying serious attention to their numbers, the census-estimated populations have sunk as low as 334 birds. Some attribute the low to the hurricane of 1973 that rampaged through their winter tropical territories, but a harder reason to face is the more likely one of habitat loss. A short-term boost in the bird's population, interestingly enough, is thought to have occurred in the aftermath of the first "big cut" of lumbering that came through middle America in the late 1800s. Although periodic openings had historically been created by natural fires, the extent of cleared land and mad slash fires that followed the big cut may have given rise to an unprecedented (albeit temporary) increase in the kind of dog's hair jack pine stands that the Kirtland's needs for nesting.

Their vulnerability is doled out in good measure: the parasitism of their nests by cowbirds who supplant the warbler eggs with their own, the dangers inherent in the long annual migrations from which only 40 percent are thought to return, and a narrowly defined breeding habitat that is this ball suspended in the air, a community in perpetual adolescence. For when the trees grow high and large enough that their shade kills the plants below them, the warblers must move on to a place that gives better concealment to their ground nests. If such a place does not exist, then neither will the Kirtland's warbler.

I tell myself that it's fine if I don't see or hear one. After all, it's midsummer. The males have less reason to sing, and both the male and female have more reason to be secretive. Their territories require less work to defend. The nestlings of their first, and often only, brood of the season will already have fledged. If a second brood is under way, the female may be incubating eggs in her soft nest of grass and deer hair, hidden beneath the prairie grasses and blueberries. In any case, they have better things to do than be seen.

From this vantage along a national forest road (it is illegal to enter the forest during the breeding season, since the Kirtland's is federally listed as an endangered species), I train my binoculars on the jack pine cones clinging to the tips of the trees. In the poetry of adaptation, the warblers' fledglings are said to resemble nothing so much as the cone of a jack pine. I see only pine cones. That is, as far as I can tell. Among them may be a great impersonation of a pine cone.

There is a sexiness about rare animals that can blind you to what is around them. Everything in the vicinity is measured as good or bad relative to its known influence on the creature whose population is in trouble. The result is a skewed and partial vision of what is a much more complex system made not only of the known, but more important, of the *unknown* elements that make it work.

This particular stand of jack pines is here, not from fire, but because it was planted with the sole intent of creating warbler habitat that had diminished because of our modern habit of putting out the fires that lightning starts. Fortunately, these managed stands of jack pine have proved to be acceptable breeding sites for the Kirtland's. But those who stand, as I am today, in hopes of seeing one of "the 1,266," are not only in the presence of the Kirtland's summer range. Beyond this manmade dog's hair stand with its precious dwellers is the larger community within which it exists.

Before the era of forests as crops, this part of Michigan was a blend of dry sand prairie and pine barrens. It was not the solid green curtain of forest that many have now come to consider synonymous with natural beauty. It had its own brand of beauty, a stark and wind-whittled eloquence expressed not only in trees, but also in bluestem and rice grass, Indian grass, the yellow flowers of the pale agoseris, the bright green fans of rough fescue grass, and the Hill's thistle that would spend four years preparing for its one white blossom. With them were the animals of open country: the vesper sparrow, the clay-colored sparrow, the hog-nosed snake, the coyote, the badger.

Apart from management for the Kirtland's warbler, work is also under way at the Huron National Forest to clear some tracts of land to allow this other rare element to survive: in a word, the element of "openness." Each season following a clearing reveals more and more of the historic native plants trying to recolonize after two hundred years of suppression, reveling in the so-called poor soils that, for them, are the soils of home.

The song is clear and loud. I see the male with his deep blue back, yellow streaks on his breast. His body shakes with the effort of every note. The labels and numbers fall away. I can tell in a glance that he does not know he is endangered. He knows only that his job is to sing, this day, from the top of that young jack pine. His beak is open, full of the sky behind him.

ECO-REGION V

Northern Lake-influenced Upper Michigan and Wisconsin

Lay of the Land

Most of Eco-region V is characterized by low relief, dominated by flat clay and sand glacial lake-plain topography. Along present-day coasts and the now inland sites of historic lake plains and shores are found such features as parabolic dune fields, deltas, and undulating ridge and swale topography. The Niagaran Escarpment, a resistant ridge of dolomite and limestone bedrock, underlies Wisconsin's Door Peninsula and curves along the entire southern length of Michigan's Upper Peninsula.

Native Communities

Similar to the northern half of Michigan's Lower Peninsula, communities native to this region include northern hardwood forest, jack pine barrens, white and red pine forest, and hardwood-conifer swamp. Hemlocks and white and red pines have largely been lost to logging for timber and, in the case of hemlocks, for the tannins derived from their bark. The region retains fine peatlands in the eastern lowlands of the Upper Peninsula, extensive forested swamps (dominated variously by white cedar or black spruce and tamarack), as well as Great Lakes coastal marshes and estuaries in embayments and protected coves. Alternating ridges and swales along shorelines are host, respectively, to upland forest and lowland sedge meadow/emergent marsh. Globally, rare grassland communities are found associated with exposures of the Niagaran Escarpment.

Animations

• A cool microclimate is created in a strip along the western shore of the Door Peninsula in summer. Winds coming out of the southwest blow the warm surface waters of Lake Michigan westward, causing an upwelling of deeper, colder water along the shoreline. The conditions support the unusual occurrence of plants such as bearberry and arctic primrose, traditionally found as much as two hundred miles to the north. The cold upwellings are also favored by fishermen, since trout and salmon may be found at much shallower depths in the upwellings than they would typically be found in summer.

• Entire beaches of Michigan's Upper Peninsula are formed from fossils of sea creatures that swam in Paleozoic seas more than four hundred thousand years ago.

• A series of waterfalls is found along the length of the northern Upper Peninsula where rivers heading toward Lake Superior encounter a ledge formed of ancient Cambrian sandstones. Among the falls are Tahquamenon, Miners, and Laughing Whitefish.

• In all the world, only rain forests and saltwater marshes are considered to be more biologically productive than Great Lakes coastal marshes. At least two critical marsh systems exist within this region: the marshes between the Straits of Mackinac and the Les Cheneaux Islands, and the marshes along the St. Marys River between Sault Ste. Marie and the village of Detour. While these marshes occur in areas sheltered from direct wave action, they rely on the fluctuations in water level they receive by their association with the Great Lakes.

• There have been numerous reports of cougar sightings, including observations of adult cougars with young, in the central part of this region. At least one sighting has been confirmed as recently as 1985.

Border Justifications

The region is distinguished from Eco-region IV by the physical barrier of the Straits of Mackinac dividing Michigan's Upper and Lower Peninsulas. The western border with Eco-region VI is located along a line of transition in bedrock from Eco-region V's Paleozoic sedimentary bedrock of sandstone, limestone, shale, and dolomite, to Eco-region VI's older Precambrian "Canadian Shield" bedrock, which includes sedimentary conglomerates and shales, but is dominated by igneous and metamorphic rock types such as gneiss, basalt, and granites. The border between Eco-regions II and V is a reflection of differences between soils developed under Eco-region II's savannas and grasslands, and Eco-region V's northern forest of hardwoods and conifers. The soil of Eco-region V includes an acidic and nutrient-poor *spodic* horizon, formed when water percolates through needle duff and picks up acids that leach mineral salts from the top layer of soil into deeper layers.

Drummond Island and Maxton Plains

July

They paved paradise . . .
(Joni Mitchell, "Big Yellow Taxi")

A person benefits from knowing certain things when in the Upper Peninsula of Michigan. First, the many signs reading "Best Pasties in the Upper Peninsula" generally do not refer to certain clothing worn by exotic dancers, but rather to a kind of meat pie with rutabagas. Knowing this helps one decide how many to order. Second, Mackinac does not rhyme with Cadillac. The "-nac" is pronounced "naw." Knowing this helps one to avoid being the recipient of pained looks.

It is also good to know that loons call in the early morning fog along the northern shores of Lake Huron east of the Straits of Mackinac; that rare Pitcher's thistles bloom on the sand dunes; that sturgeon and mooneyes and ciscoes still spawn in the coastal marshes; and that, for a few weeks in spring, there is little more important to a hungry migrant warbler than the hatching of midges from exposed rocky shoals in the shallows. Knowing these things helps one to live. And to let live.

Drummond Island is a short ferry trip off the eastern end of the Upper Peninsula. On the crossing I can see the vague shapes of other islands to the north, some of which belong to Canada.

The border between nations is shown on a map as a dashed line across the waters of Potagannissing Bay. In reality, the border is marked by a line of red buoys whose authority doesn't seem to mean much to the wind and waves.

When the ferry docks, I head toward the northern end of the island. Traveling from south to north on Drummond is the equivalent of growing older in geologic terms. Here at the perimeter of what is known as the Michigan Basin, the layers of bedrock curve up like bowls in a nested stack so that, traversing the surface, one encounters bands of successively older rock. The southern part of the island is analogous to the rim of the inner-most bowl of the stack, with its layer of rock formed in the Silurian period roughly 410 million years ago. The northern part of the island would represent the rim of the outermost bowl, originally a deeper and therefore older layer of rock formed during the Ordovician period roughly 450 million years ago. It takes about twenty minutes by car to traverse some 40 million years. I'm not sure if that is a good or a bad thing.

The paved road ends about three miles from the island's northern shoreline, in a Nature Conservancy natural area called Maxton Plains. After that, a road of sorts runs on for a while across bare dolomite bedrock, on which someone has creatively painted a yellow center line. Not that the road gets much use. The summer people with their houses on the Lake Huron shore mostly come by boat. In fall there are some bear hunters. But the winter people, well, there aren't many winter people.

Maxton Plains has the unusual distinction of being a natural area that could be mistaken for a neglected parking lot, complete with weeds growing up in the cracks. It's the kind of place where you wouldn't be surprised to come across a group of kids playing kick the can.

But if there were kids here, instead of playing on asphalt or concrete they would be playing on an old seafloor, part of the same escarpment of dolomite rock that arcs across the continent to form Niagara Falls. The "weeds" that they leapt over in their game would be a rich assembly of native plants *dominated* by uncommon and protected species. Their audience might include a veery or a sandhill crane nesting in the grass; or during a night game, a woodcock whirling in the skies. They would be in the midst of one of the finest known examples of a natural community that is itself rare, limited in the world to only a few sites in the United States, Canada, and along the coasts of the Baltic Sea in Scandinavia: a natural community called alvar.

Alvar is more a phenomenon than a discrete set of parts, characterized by plants that grow in thin soils over flatlands of calcium-rich (limestone or dolomite) bedrock. "Thin soils" is an understatement. On the high end of the abundance scale, there might be eight inches of soil—enough to allow for a covering of plants, given an ample supply of water. On the low end of the scale—in so-called pavement alvar—are broad exposures of bare bedrock where rooted plants must follow fracture lines in the rock in order to find any soil or moisture at all. Accordingly, alvar can wear many faces, appearing variously as a blanket of green or as a mass of rock traced by thin green lines.

At its heart a grassland, an alvar community is a very different place in May than it is in August. In spring, cold water pools on the bedrock and saturates what little soil exists, creating conditions in some ways similar to that found in tundra environments. By late summer most of the moisture is gone. Extremely arid conditions prevail that could be likened to the dry prairies of the west, and it is not uncommon for all of the plants of the alvar to "brown off" above the surface, with the exception of those that manage to find water in the rock fractures. The result of these seasonal swings, at least on Maxton Plains, is a quirky combination of plants that are traditionally found in dry prairie environments—prairie dropseed, prairie smoke—and plants generally considered to be boreal (northern) species, such as the downy oat grass and bulrush sedge. Botanically speaking, alvar is an odd duck.

I am told that I am too early. The first week in July is coming to its close. In another week or so, the flowering plants are expected to be in full bloom, and the alvar will be consumed by color that could rival any mountain meadow. Already I can see its beginnings in a spatter of indigo blue harebells and a few orange and yellow "Indian paintbrushes" flaring up in the prairie grasses. Looking out on the landscape's subtle beauty, I could admit to feeling many things; early is not among them.

I wade out into the grassland as into a shallow green sea. For the most part, the capriciousness of the water supply in alvar communities puts them out of bounds to trees and shrubs. Seasonal drought has taken on the role of fire in prairies, the role of preserving grassland by keeping the forest at bay. Not that the seed source isn't here. White spruce, white cedar, trembling aspen, and balsam fir all crowd at the edges of the clearing. These are trees that could typically cover the land at this latitude, given the right conditions. Year after year they send their offspring out to give it a try on the thin soils of the alvar. Only those that manage to land in the vicinity of a fracture line stand a chance at establishing, and of those, few other than the juniper shrubs seem to be able to endure more than a season or two. In this way is the alvar sustained. And with it, the tawny crescentspot butterfly and all the other rare species for whom trees would only get in the way of making a living.

Coming upon a patch of pavement alvar, I find something eminently reassuring about the sight of the bright green lines of native grasses spurting up through the cracks of the dolomite. It has nothing to do with adversity. It has to do with continuity, with the acknowledgment that the present is just the past with some time on its hands. These plants are new lives rooted in old lives, for the rock itself is an accumulation of ancient animals, its calcium carbonate born of coral and the shells of sea creatures that once fed and swam, and in their own way, strived.

On a small island east of one place and west of another, south of one place and north of another, there exists something whole, something unbroken. It is a good thing to know.

Door Peninsula

July

The roads of the peninsula are patrolled from above by beings whose heads are composed mainly of eyes — red eyes, black eyes, milky white eyes, eyes of brilliant green. Eyes that have been refined over a span of more than 220 million years. Eyes that are able to detect the slightest of motions, less able to distinguish detail.

It could be that they perceive the linear reflective surface of the road as a river and that they are claiming it as territory. A river would accept the eggs of the beings, deposited freely in the flowing waters or inserted with precision into openings razored into the living tissues of shoreline plants. From the eggs, in time, would hatch into the river the armored forms of their young, which — until four centuries ago — had been known as "water lizards." Then it came to be known that they were the offspring of the beings in the sky: the offspring of dragonflies.

Sometimes the dragonflies that fly above the roads will dip down to touch a tip of abdomen to pavement, just as they will touch the water of a river, as if to confirm what their eyes ask them to believe. Discerning, if not between river and road, then at least between right and not right.

≋

Of a summer morning, the Door Peninsula takes its usual place between Green Bay and the open waters of Lake Michigan. The peninsula is a north-pointing strip of land ninety miles long. Its width varies between four and eighteen miles and is traversed by lowland valleys that follow joints in the limestone bedrock. The joints run parallel in the bedrock, trending northwest to southeast like a ladder laid along the length of the peninsula, its rungs knocked aslant. And so it is that the valleys, too, are oriented. Some are buried by glacial till. Others hold cedar swamps, wetlands, or streams, and open out into bays on either coast that scallop the margin of the peninsula.

The most deeply incised of the lowlands aligned along the joints have become water-filled channels that break the finger of land into segments. There is the break at Sturgeon Bay, incomplete until shipping interests gnawed through the last mile of land by which the peninsula was attached to its base. And then the natural break at Porte des Morts, French for "Door of Death": a wind-ridden, boat-eating strait that separates the peninsula proper from Washington Island at its tip.

As a whole, the surface of the peninsula is inclined toward Lake Michigan. Its underlying layers of bedrock begin as high escarpments along the Green Bay shores and descend eastward at a rate of forty-five feet per mile.

On the relatively lower east coast, Lake Michigan whittles and builds, gives and takes. In one place it undercuts the bedrock to make caves into which its waves may then boom and spit, making the ground tremble at each strike. In another, the lake brings sand that southerly winds build into dunes. In still another are found the stair-step of ridges and wet swales left in a band along the shore at each successive drop in elevation of the lake through its history.

Temperature of lake, direction of wind, height of land: these are among the forces that pattern the natural communities of the peninsula. Where westerly winds blow the warm surface waters of Lake Michigan eastward, upwellings of cold water from the depths create a microclimate near shore in which spruces and firs flourish, with orchids at their feet. Elsewhere, the peninsula's uplands support sugar maple, beech, some white pine, and hemlock, while in the lowlands are white cedar, hemlock, balsam fir.

Over it all is the light, such phenomenal light. The wild country as well as the tended landscapes of rolling farm fields and orchards of cherry trees all glow as if lit from within, caught in the cross fire of sun banking off the surrounding waters. It is a quality of light I believe would beckon to the brushes of van Gogh.

It was what I noticed first when I had crossed the bridge at Sturgeon Bay the day before, when I was beyond the outskirts of the city and into the open countryside of the peninsula. That lovely light. While still in full sun, I had watched as a squall line came across Green Bay. I could see the rain in the distance, saw the thunderhead lifting bright nets of lightning from the water. It came nearer, overtaking the land. I was headed north. The storm was headed east. I tried to guess how soon we would meet. But we never did. It was gone, already out over Lake Michigan, by the time I crossed its path. Instead of rain and lightning I found dripping cherries, steaming roads, the scent of wet earth.

The rain that fell from those clouds would have sunk with speed into the groundwaters of the peninsula. There would not have been the slow percolation that might have purified it of toxins, for the peninsula is underlain by dolomite, a type of limestone. It is in the nature of limestone to be pocked by caverns and fissures that would have offered swift passage to the water. Whatever the rain had gathered from the air, whatever it washed from the roads and orchards, the fields and topsoil, would have gone with it into the shallow aquifers, and thus, directly into the springs that feed the rivers. Until twenty years ago, it also went into the wells. Now law requires that wells be drilled deeper. Steel pipes are put 170 feet down and then the wells dug still farther beyond the pipe to reach aquifers at 200, even 250, feet.

perfect, the seams precise. There are the curves that held the immature dragonfly's abdomen, and the plates of the gills it used to breathe during the years when it lived underwater. Holding it gingerly, so as not to break it, I lift it up so that I can peer through the goggles that once covered its eyes.

ECO-REGION VI

Northern Continental Michigan, Wisconsin, and Minnesota: Superior Shores, Hardwood Forests, Bedrock Cliffs

Lay of the Land

The topography of Eco-region VI is dominated by moraines formed beneath and at the margins of glacial ice. Other features include a high plateau of volcanic and metamorphic origin that rises nearly 1,400 feet above Lake Superior, broad outwash plains of drift deposited by glacial meltwater streams, and lakes in depressions formed by glacial scouring of bedrock or by the melting of isolated blocks of glacial ice. Islands, dunes, sand spits, and river deltas are important features along the region's nearly two hundred miles of Lake Superior shoreline. Clay beds of ancestral lake plains extend inland as far as thirty miles from the present-day shoreline.

Native Communities

The primary forest community of this region is its outstanding northern hardwood forest of sugar maple, birch, eastern hemlock, basswood, and scattered white pine, of which significant virgin tracts remain. Outwash plains support barrens communities of jack pine and northern pin oak, while undisturbed areas of the glacial lake plains harbor Great Lakes marshes, dune complexes, and hardwood-conifer swamps, as well as white pine–hemlock forest and spruce-fir forest. Exposed bedrock knobs are host to red oaks and red and white pine. Many areas that underwent intense slash fires following settlement-era logging have grown back to aspen–paper birch forests.

Animations

• The Lake Superior basin is still rebounding from the weight of the last glacial ice that departed the region roughly ten thousand years ago. This *isostatic uplift* is causing the northeastern corner of the basin (in the Michipicoten area) to rise at a rate of nineteen inches per century relative to the southwestern corner of the basin at Duluth. Corresponding uplift of the lake's outlet at the St. Marys River is causing water levels to climb slowly but steadily all along the lake's southwestern shores in Michigan, Wisconsin, and Minnesota.

• The region's extensive northern hardwood forests are considered critical breeding grounds for migratory songbirds. Large, unfragmented forests such as those of the Porcupine, Sylvania, and Huron Mountain wilderness areas are thought to be reservoirs from which declining populations of these birds in surrounding areas are replenished.

• Chemicals banned in the United States but still sold by U.S. companies to other countries may well be finding their way back home via global air circulation patterns. Fish taken from inland lakes of remote and wild Isle Royale, Lake Superior's largest island, have been found to have extremely high levels of airborne toxins, including DDT.

• Lake-effect snows from moist air coming off Lake Superior are greatest at inland points of high elevation along the shore. This increased precipitation related to a rise in elevation is referred to as "orographic" or "orthographic" precipitation.

• The Menomonie Indian Reservation in the east-central part of the region has received international attention as a model for forestry and timber management. Although Native Americans have actively harvested timber from these lands for centuries, their management to cull the poorest rather than the strongest trees, a long rotation time between harvests, and their philosophy of striving to maintain a stand composition of mixed species and ages, have meant that their forests have retained a remarkable degree of integrity as natural communities.

Border Justification

The extent of the region's Precambrian bedrock defines its eastern and southern borders. To the west, its border with Eco-region VII is distinguished by a shift from the dominance of northern hardwoods on moist upland sites in Eco-region VI to a dominance by conifers on similar sites in Eco-region VII. The shift is thought to be due, in part, to an increase in drought-related fires in Eco-region VII, conditions that favor fire-dependent conifers over northern hardwoods.

Kakagon/Bad River Sloughs

July

It is more than rice and walleye. It is the place where we go
to heal. I have never gone out there and not felt better
afterward than I did before I went. I have to assume
it does the same for the animals.
(Joe Dan Rose, Bad River Band
of Lake Superior Chippewa Indians)

Don't believe me when I say that the Kakagon/Bad River Sloughs
look like the Everglades. It only means that I have not spent
enough time here. If I had, I am sure that I would say instead that
the Kakagon/Bad River Sloughs look like the Kakagon/Bad River
Sloughs.

But I can tell you this. The rivers have no shores. They move
without argument through a yielding sea of vegetation, through
plants without spines, plants that sway slightly even on a wind-
less day. The line of sight is long. It is a world of horizontal layers. Blue, green, blue. Water, manoomin, sky.

Manoomin is the Chippewa word for wild rice. It edges the
Bad and Kakagon Rivers as they flow through the sloughs in this
pristine wetland along Lake Superior's south shore. Now, in
mid-July, the manoomin glows a pale green, reaching about two
feet above the surface of the water. Some stems hold purple and

white flowers. Others have already set seed, but none is ripe. It is only a matter of time—three, maybe four weeks. When the plants turn the color of straw.

The red-winged blackbirds will know when it is ready. So, too, will the people of the Bad River Band, whose land this is, and who will come in canoes to harvest some of the rice for their families' winter tables. For them it must, I think, taste of pride: pride that the Band's steady loyalty to the land has kept this landscape wild enough for the rice, and for the tangle of lives—human and otherwise—that are bound to the 16,000-acre sloughs. How fine it must feel to be the answer to the question, "Why isn't this place gone?"

Popcorn clouds cluster at the horizon. A great blue heron lifts off from the grasses and into languid flight, its wings cupping the air, long legs following behind as afterthoughts. I see with my own eyes, and I see with the eyes of those who have been here before me, through the stories they've told. There is the black bear swimming across the channel of the Kakagon. The moose and deer stepping stilt-legged across the sedge meadows, their trails fanning out behind them. There are the mallards, teal, bluebills, and canvasbacks that coat the waterways not long after ice-out in spring. And in late April and early May, before the new growth of grasses has broken above the waterline to define the river channels, there are the walleye and then the sturgeon coming in from the lake to spawn.

I ride at the bow of a flatbottom boat, here in the sloughs with the kind permission of the Bad River Band, a matter of both courtesy and law. The boat glides on the calm water, riding so smoothly it seems not to move at all, as though instead it is the landscape that is drifting by the boat. Lilies at the water's edge blend into beds of wild rice and pickerel weed. Ragtag stands of red maple and black ash straddle the low natural levees that parallel the river channel. Off to the east, the tips of scattered tamarack and spruce mark the presence of a bog.

Ed Leoso sits behind me with his hand on the tiller. He is a fisheries technologist for the reservation and works, among other things, on a project that tries to increase the reproductive success of native walleye by collecting and rearing their eggs, then releasing the fry back into Superior. The boat motor is running at a level just loud enough that I abandon the idea of conversation — a good strategy, perhaps, on Ed's part. He interrupts our silence only to call out the names of the tributary streams we meet as we follow the Kakagon toward Chequamegon Bay. "Wood Creek," he calls, pointing. Then, a few minutes later, "Beartrap."

The sloughs are a gathering place for waters. One by one, hundreds of named and unnamed creeks and rivers splice together on their descent toward the lake, draining a thousand-square-mile watershed. The last tributaries join in on the broad clay plain of the sloughs. Finally only two rivers — the Kakagon and the Bad — are left to carry the collective waters into Superior. The smaller Kakagon meanders in wide loops through the sloughs and drains westward into the bay protected by the long sand-spit arm of Chequamegon Point. The Bad flows north and east into the open lake.

In spring floods and on windy days, these rivers run blood red, the color of the clay sediment picked up by the currents. Whatever the river waters find along their route they bring to the sloughs. Their pH, their load of sediment, the quality and volume of water they carry: all are felt sharply at these lower reaches of the watershed. But the secret to the sloughs is that the flow of water in these channels changes direction periodically throughout the day. Like the rivers along ocean coasts, each river is really two rivers — the one that flows out to Superior and the one that flows inland. The Great Lakes, it would seem, have a kind of tide.

Never mind the moon. Although the lake does experience a small lunar tide (as does, for that matter, a cup of tea), Superior has its own reasons to move. Weather systems traveling across the surface of the lake bring with them two great forces that influence the waters: wind and barometric pressure. While these forces do not change the actual volume of water, they change its *position* by imposing a temporary tilt of the lake in its basin. High pressure "pushing down" on one part of the lake will create a corresponding rise of water elsewhere. Steady winds will pile water on one shore, bringing about an associated drop in elevation on the lee shore.

The lake naturally seeks to regain its level surface when the winds eventually cease or the high pressure system moves on. Released of the force that held them there, the areas of high water rush back toward the areas of low water. If there were no new forces, the waters would simply rock from shore to shore in their basin, much like water in a glass after it's been bumped, until the energy was spent and the lake was once again calm. But Superior has a very large surface area upon which weather systems can play. The lake will often still be moving from a past weather event when another comes to set it rocking anew.

The term *seiche* (pronounced "saysh") is used to describe these movements of large lakes in their basins as they recover from a wind or pressure event. A seiche can mean that the water level unobtrusively climbs up and down a few inches at the base of the blades of grass in a coastal marsh. Or it can be a beast of a very different sort, one that races across an otherwise calm lake with thirteen-foot waves that overturn boats and swallow the lives of the unsuspecting on docks and shores.

For reasons not fully understood, seiches seem to be experienced in the Kakagon/Bad River Sloughs as a surprisingly regular —that is, *harmonic*—pattern of ebb and flow. Failing storms and other wild weather that would override their influence, "high tide" and "low tide" tend to alternate in the sloughs at roughly two-hour intervals.

Here, the water level change between extremes is generally only four or five inches. Yet the movement is enough to enrich both the sloughs and the lake beyond measure. The influx of water from the lake into the sloughs allows for periods of still water during the "turning of the tides" when nutrient-rich sediments are able to settle out, thus fostering the growth of the wild rice and the other lives of the marsh. In turn, the river water flowing into the lake provides the main source of organic food to the coastal communities of fish and other aquatic life.

The sloughs have been characterized as a freshwater estuary. The expression is somewhat of an oxymoron, given the traditional definition of an estuary as a region where freshwater and saltwater intermingle. Still, there is an undeniable kinship between Great Lakes river delta regions such as the Kakagon/Bad River Sloughs and their counterparts on ocean coasts. Both are dynamic environments based on tidelike interplay. Both embody the precarious juxtaposition of three traits: biological diversity, rarity, and fragility. Both are endlessly engaged in the fine art of reinventing themselves.

The boat rounds a bend. A hen mallard skitters along the waterline, head pumping in concert with the urgent paddling of her feet. She looks to be in perfect plumage, and I wonder why she doesn't fly. Then her three young reasons emerge from between the blades of wild rice behind her.

The noon sun of July beats full upon the sloughs. This is the time of year when, in the farms a hundred or so miles to the south, the corn can gain as much as five inches of height in a single day. If a person were so inclined, he or she could stand in the fields and listen to the leaves unwhorling from the stalks, literally hear the corn grow. For Superior the season means instead the onset of the high water season, the long-delayed response to spring runoff. At the other end of the spectrum is the lake's low water season in March and April.

Like the child who sleeps most soundly in the midst of commotion, the sloughs find their own brand of stability right in the midst of what might appear to us to be instability. The sloughs thrive — no, *rely* — on the changes in water level that occur daily, annually and even on five-hundred-year cycles. Flooding in high water seasons keeps trees and other woody vegetation from invading, preserving the diversity of wetland communities. Low water seasons offer the dry seedbed that many marsh plants require to establish. What seems to be chaos becomes an ordered and breathtakingly intricate pattern when interpreted at a broader scale.

I watch as the Kakagon widens to enter the bay. Directly ahead is the nub of Oak Point. Out of sight around its curve is the twelve-mile-long sand spit of Chequamegon Point, where the first of what Peter Matthiessen called the "wind birds" will already have begun to stop in for a rest as they filter south on migration: dunlins and dowitchers, terns and sandpipers, godwits and knots.

At the very tip of Chequamegon Point is Long Island, part of the Apostle Islands National Lakeshore and the last known Wisconsin nesting site of the endangered piping plover. Truth be told, Long Island is not an island. It stopped being an island in 1975 when the space between the island and the mainland was filled with sand by the same storm that downed the legendary ore tanker, the *Edmund Fitzgerald*. No one has since bothered to change the maps. There is bound to be another storm, sooner or later, that will take the sand away again.

Ed pushes the tiller to bring the bow of the boat around in a broad swing to head back into the sloughs. Somewhere here, Lake Superior ends and the sloughs begin. Somewhere else, ahead of us, the sloughs end and the land begins. I couldn't say just where.

Keweenaw Peninsula

Late August, Early September

End of the World 2 miles
Houghton, MI 4 miles
(Road sign depicted on a postcard designed
by students at Michigan Tech)

Free Piece of Native Copper with Gas Purchase
(Sign in a gas station window, Copper Harbor, Michigan)

It is late August when the bats begin to filter into the abandoned copper mines of Michigan's Keweenaw Peninsula. Through the summer there have always been a few. Now they come in earnest into the tunnels where miners from England once tore penny-bright veins of copper from milky calcite, working the rock by candlelight with star drills and sledge hammers. When the first hard frost knocks most of the flying insects from the night skies, the ceiling rock of the mines will be trimmed with pendulous clusters of "little browns." By October the Delaware Mine alone will shelter two thousand. They'll sleep through the winter in the steady forty-five degrees of the hundred-foot-deep tunnels, kept warm not by the radiant heat of the sun, but by the earth's own molten heart. The darkness will be absolute.

The Keweenaw: northernmost reach of Michigan, it is a bull's horn of land extending off the top of Michigan's Upper Peninsula, surrounded on three sides by Lake Superior. A peninsula's peninsula.

Depending on your perspective, the Keweenaw is mining country, logging country, a refuge for famed but poignantly small groves of lofty old white pines, a good place to hang head-down for the winter, home of "The Jam Lady" and her wild thimbleberry jam, or a shoreline to rest on during spring migration before the long flight across Lake Superior. Today, on a ridge above Eagle Harbor, it is mostly rock, rain, and wind. Gray veils of clouds drift inland along the forested valley floor some six hundred feet below. The rain has broken all ties with gravity; it slices across the sky in horizontal sheets and flies with gusts of wind straight up the rock face of the cliff. I couldn't begin to think how a person might position an umbrella, not that I have one.

I am glad for the rain. Glad for the wind. Here at the margins of Lake Superior, in this season, anything else would be intermission. The grace period that the lake offers the Keweenaw in summer is approaching its end. When the lake is colder than the air above it, as it is through the summer months, wild weather is rare on the peninsula. Thunderstorms coming on the prevailing westerlies tend to die out when the stable air over the lake refuses to feed their updrafts. But now, as August gives way to September and on into autumn, a new set of conditions comes into play. The jet stream dips down from Canada with its cyclones. The winds begin to shift more northwesterly. The lake has held on to some of the heat it gathered when the days were long, but the temperature of the air above it has become progressively colder. Storms are now strengthened as the warmer air immediately above the lake rises. Rain, drizzle, and fog are more likely. The seeds have been planted for the Keweenaw's notorious snows of up to thirty-two feet in a season.

I watch a raven soar out over the precipice as if the distance to the ground were irrelevant, and I wish, as I have wished before, that I were practiced in the shape-shifting of the Navajo and the Yaqui. My intentions would not be malevolent. I would steal no souls. With a Jimmy Durante beak, tail feathers tapering to a spade point, ebony wings, I would circle this ridge again and again just to feel the land fall away beneath me. If you asked the distance to the ground, for once I would not know.

Standing is nearly impossible. I drop to my knees at the bald summit of the cliff, ringed by trees that all their lives have done the same. These are the "krummholtz," the twisted ones. Trees not only exposed to storm winds but also battered by the summer night winds that race up the peninsula toward the lake. Like the elfin forests of the Neotropics, the high alpine communities of the mountains, and the salt-sprayed vegetation along seacoasts, they are left with faces wizened beyond their years, and without the chance of ever growing tall. White pines and cedars crawl like shrubs. Spruce and fir are reduced to green mats and bare stems. There are dwarfed, flat-topped red oaks no closer to the sky at the age of seventy than they were at the age of twenty. A juniper turns around on its stem as though looking northward to the place where its youth went with the wind.

Beneath me, forming the peninsula, are layers of rock that harken from what is known as the Keweenawan period, a time late in the Precambrian era between 600 million and 1,640 million years ago. The period was witness to the rise and subsequent slow deterioration of the Penokean Mountains, whose high peaks once broke the skyline from present-day South Dakota eastward to New York State and up into Quebec. Formed from the debris of the range's erosion are many of the peninsula's sedimentary rocks: fine sandstones, shales, and the fused jumble of rock, sand, and clay known as conglomerate. These same rocks combine with rock layers of volcanic origin to line Superior's basin. They are depressed under the lake at the site of a past continental rift, and tilted up at the edges to form a series of parallel ridges, on one of which I now kneel.

To the north, the east, the west, the land rolls gradually down and away through the mist toward Superior, hesitating for a moment at the waterline to shed its clothing of trees before diving headlong into the lake. The rocks of the Keweenawan series will rise again briefly on Isle Royale, more than four hours by boat off the tip-end of the peninsula. They will also rise in the escarpments of the Porcupine Mountains Wilderness, eighty-five miles to the southwest as the raven flies, across Misery Bay.

Cars *shoosh* by on a narrow road below, windshield wipers waving. No reason to stop. No vista on such a day as this. If someone were to look up and see me here, kneeling on the ridge, they might wonder at the woman staring off blindly into the soft gray cloth of clouds. They would have no way of knowing that, as the woman on the ridge looks toward the southwest, in her mind's eye she has traveled those eighty-five miles across the bay to the wide base of the peninsula. They would not know that she is, this moment, deep in the Porkies' 35,000-acre old-growth northern hardwood forest of yellow birch, maple, and hemlock.

The ground under her feet is soft, covered in the damp fire-resistant duff of hemlock needles. She moves among mushrooms as colorful and delicate as flowers, smiling at the quiet threat of a poisonous amanita— *the destroying angel*—with its pure white cap and skirted stem.

The woman sees that, here in the Porcupine Mountains, as on the ridge above Eagle Harbor, wind has been the purveyor of change. An oval pool of young trees, all of a size, cover a hillside. They have grown up from the ruins of a thunderstorm downburst that decades earlier had laid flat a broad swath of forest with hurricane-force winds. Beyond them she finds a valley that may well have escaped major disturbance for millennia. Change has been more subtle here, measured in small steps accumulated over time. A single giant yellow birch downed by wind has become a nursery log for a dozen maple and hemlock saplings. They mine its rotting length for nutrients and moisture, and pile in brief freedom from competition atop the tip-up mound of earth that the big tree dragged up as it fell. She sees the future of the young saplings mirrored around them, in scattered grown trees that are poised as if to walk, each with roots cupping empty air like the legs of a crab, defining the space where their nursery log or tip-up mound once was.

The people in the passing cars would not, could not, know that as the woman on the ridge turns her head to look instead to the east toward the tapering end of the peninsula, in her mind's eye she walks now on a muddy path down a series of terraces toward Horseshoe Harbor. She passes through a fog-misted forest of cedars and spruce and fir. The woman stops once to pick a ripe thimbleberry and then again, to watch an ovenbird ruffle its crown feathers and lay them back down. She steps onto the red cobble bedrock beach of the harbor. Considers a sky that knows the feel of a peregrine's wings. Considers the calm curl of the harbor, protected from Superior's wild west by a great wedge of conglomerate rock, mother to the cobbles of the beach. She picks up one of the rounded Precambrian stones, surprised that six hundred million years could weigh so little in her hand.

On a cloudy day, on the Keweenaw, you can see just about forever.

ECO-REGION VII

Northern Minnesota: Subboreal Forest, Peatlands, Border Lakes, Superior Highlands

Lay of the Land

The entire area of Eco-region VII is characterized by Wisconsin-era glacial drift and landforms, including rolling to steep ground- and end-moraine ridges, glacial lake plains, and sandy outwash plains. The bedrock of the Canadian Shield is at or near the surface throughout much of the northern portion of the region. Major features of relief include the Giant's Range—a ridge of granite associated with the iron-rich Mesabi Range, which rises four hundred feet above adjacent plains—and the Superior Highlands, a ridge of Keweenawan volcanics and conglomerate following the Lake Superior shoreline.

Native Communities

Key communities include upland conifer forests, mixed conifer-hardwood forests, peatlands, and conifer swamps. Northern hardwood forests are found in the more southerly reaches, tapering off northward except in areas along Superior protected from fire and frost. Northernmost reaches of the region grade toward spruce-fir forests, akin in composition to the boreal forests of Canada. Poorly drained lake plains of Glacial Lake Agassiz are characterized by peatlands and spruce and tamarack swamps. Jack, red, and white pine were historically found on outwash plains and sandy moraines, with jack pine found in the more fire-prone sites. Logging has dramatically increased the percentage of white birch and aspen forest throughout the region. Thousands of lakes occupy ice-block depressions and glacially scoured bedrock.

Animations

• Although prevailing winds generally push lake air masses eastward, Lake Superior still offers Minnesota's near-shore regions up to ten more inches of snow and a growing season ten days longer than areas five miles inland at the same latitude.

• The region's forests represent the last significant clawhold on the "Lower 48" for the eastern timber wolf. An average pack of six individuals can require an area as great as 150 square miles from which to obtain their food.

• Fires play as intrinsic a role in maintaining northern forest communities as prairies. Areas that experience periodic intense fires tend to regenerate to jack pine and spruce stands. Where mature trees are available as a seed source, less frequent low-intensity fires provide a competition-free seedbed that nurtures the growth of red and white pine stands.

• Ancient pollen found in lakebed sediments is used by researchers to interpret the migrations of tree species into the region following the retreat of glaciers. Findings suggest that jack pine and black spruce were first to follow on the heels of the ice, coming up from ice-free areas in the south and southwest to arrive in northern Minnesota roughly ten thousand years ago.

White pine is considered to have emigrated from the east coast and/or Appalachians to arrive a good three thousand years later.

• Female black bears with cubs routinely make their spring and summer beds at the base of mature white pines, theoretically because their cubs are able to readily climb the bark of the white pine out of danger at the mother's "woof" of warning. A combination of logging pressure, blister rust, and overbrowsing of saplings by deer currently threatens the status of white pines in the north woods.

Border Justifications

This region is distinguished from Eco-region VI, to the east, by lower annual precipitation and more severe summer drought. It is distinguished from Eco-region II, to the south, both by lower temperatures related to latitude and by its soils, which are derived primarily from Precambrian igneous and metamorphic bedrock. Its western border with Eco-region VIII is a reflection of the gradual shift from Eco-region VII's forests and peatlands to a community dominated by brush prairie and aspen groves, thought to be brought about in part by increasing drought.

Susie Islands
and Superior Highlands
August

There are times when people who live on the north shore of Lake Superior don't see the lake for days. Fog settles in on the water, and the freshwater lake with the largest surface area in the world—more than thirty-one thousand square miles—just disappears.

Last night, reading in the cabin, I heard a thump on the window screen. I looked up in time to see a bat spread its wings and fly away, gathering the insects drawn to my small light. Later, as I lay in darkness, came the sound of rain on the roof, mixing with the murmur of waves against the shore.

At morning's first light, I go down to the edge of the lake and stand on a rock just above the wet line that shows the highest reach of the waves. The horizon is lost in mist and low gray clouds. The forecast calls for afternoon showers and winds of fifteen to twenty knots. In the next few days, the winds are expected to grow stronger, so we will go today or not at all.

We follow the shoreline up to Wauswaugoning Bay, where a four-hundred-year-old cedar stands sentinel on Hat Point. Our destination is out beyond the circle of the bay, to the Susie Islands archipelago about a half-mile off shore. The island from which the archipelago takes its name, Susie Island, is the largest. Its 140 acres are managed by The Nature Conservancy as the Francis Lee

Jaques Memorial Preserve in honor of the life and work of the landscape artist whose dioramas grace the American Museum of Natural History in New York City and the James Ford Bell Museum of Natural History at the University of Minnesota. The other twelve islands in the archipelago are part of the Grand Portage Indian Reservation and as such are private land.

I have never been in a kayak. My friends show me how to set the foot supports, how to bend my knees and brace my legs against the sides so that the boat becomes stable. How to grip the shaft of the double-bladed paddle and twist the right wrist so that the off-set blades pull, rather than slice, the water. I imagine myself doing a roll, flipping the kayak in a complete circle and coming upright again. I imagine myself doing half a roll.

We push off through the shallow surf. A great blue heron lands on the arm of the bay and blends perfectly with the slate gray of a talus slope. A cormorant flies off to the north. The rocks on the bottom of the bay look close enough to touch with the paddle, pale and green through the water. It is, I know, just an illusion of the clarity of the lake. According to the batholithic map, those rocks are at least twenty-five feet down. Maybe as much as forty.

It is not without respect that I am here. I know that more than five hundred boats have been shipwrecked on this lake; that being a good swimmer doesn't get you very far when the average water temperature is forty degrees; that squalls can come in fast and unexpected from the northwest over the highlands. I know that the fog can drop without warning and limit your vision to the backs of your eyelids. And I know that the lake doesn't necessarily listen to the forecast. By coming here, we offer ourselves knowingly, not to the mercy, but to the vast indifference, of lake and wind and rock.

We round a bend and get our first look at the archipelago. The smallest islands are without trees; just fists of bedrock sprayed by waves and circled by the screaming herring gulls who make their nests there. Crowns of black spruce and balsam fir rise from the larger islands. Birches and cedars crouch at their edges, stunted by wind.

The Susies are a curious place, having had six thousand years in relative isolation since the rising waters of the lake cut them off from the adjacent land along the shore known as the Superior Highlands. Bathed year-round in cool and humid lake air, exposed to storms and washed by cold waves that sweep unencumbered from the south and east, they have developed a character all their own.

If ever time has been creatively invested, it has been here. Over the basaltic bedrock of the forested islands is spread an airy green cushion of sphagnum moss more than three feet deep. Hundreds of varieties of lichens—organisms formed from a partnership between algae and fungi—have grown at a rate no greater than half an inch a year into fanciful living sculptures that would be at home in a Dr. Seuss book: the antlered mounds of reindeer lichen, the "grandfather's beard" hanging in dusty veils from the lower branches of the spruces, and crustose lichens draping the shores with an orange as bright as ripe pumpkins.

Near the water's edge, glorying in the icy conditions to which they are well adapted, are also found a handful of species that are living memories of the community of plants thought to have occupied lands throughout the region at this latitude during the era of cooler climate immediately following the glaciers. Outside of the Susie Islands Archipelago, these plants—including northern eyebright, Norwegian witlow grass, and purple crowberry—are now more typically found hundreds of miles to the north and up into the arctic tundra.

We paddle close to an island and find a purple crowberry shrub tucked into a crevice in the rock. Wet with spray, its black berries shine like the intelligent eyes of crows. Aside from lichens, the rest of the rock face is bare. The solitude of the hardy.

The clouds break and the lake is suddenly lit with the sun's blue-white fire. On the shoreline, waterstriders skate on bright pools left on the ledges by last night's rain. Dragonflies trace their territories. A spotted sandpiper skitters back and forth at the edge of the waves. It is a world complete in and of itself, best left without knowing the pressure of a footfall.

I trail behind on the paddle back. The lake is building into easy rolling swells that come up under the kayaks, lift us up, and then drop us into the troughs of the waves. With each rise I catch glimpses of the mainland ahead: the Superior Highlands making their climb out of the lake.

Although still too far away to make out any detail in the mass of green, I consider what is held in these lands scrolling along the north shore as if I held them in the palms of my hands. Rare carnivorous butterworts cling to the cliffs beside the lake, curling their leaves to entrap insects. Dozens of streams cascade through narrow gorges to replenish the waters and bring nutrients essential to the near-shore communities of life in Superior. There are stands of red oaks for whose fat-rich acorns the black bears make autumn pilgrimages of as much as sixty miles. Pockets of northern hardwoods — maples, basswood, yellow birch — grow on the slopes above the village of Lutsen, surviving by virtue of the site's lake-moderated climate and relative resistance to fire. Near Two Harbors, a stand of regal old-growth white pines rises from abandoned glacial lake plains, while elsewhere along the shores are bedrock cobble beaches made of stones rounded and smoothed by waves that came steadily to shore before there were people on earth.

I see the highland ridges as a migrating hawk — no, as *hundreds and thousands* of migrating hawks see them each autumn. Looked upon from above. Checked for accuracy against the etched map of instinct.

Once again, the boulders loom in the green underwater light

of Wauswaugoning Bay. But I no longer need to think about turning the paddle shaft, or to watch the blade enter the water, for the motion has become familiar enough to forget. Nor do I need to turn around to see the islands behind me, or raise my eyes into the highlands, for they are within me, familiar enough to remember.

Sand Lake Peatland

October

It was in Denmark, 1952. Saturday. The men were out in a bog near the village, cutting squares of peat to use as fuel to heat their homes, when they found the body. They cleared away the cold and sodden peat from around the head and shoulders, and then the legs, until the man was completely free from the bog. They saw that the body was hardly decomposed; he could not have been dead for long. A woman from the village said that she could identify the man. It was Red Christian, she said, a peat cutter she had known who had disappeared without a trace. Judging from the man's teeth, investigators estimated that he had been about thirty years old when he died. Then they ran a test, called radio-carbon dating, to find out how long ago he had died, how long he might have lain in the bog. The results of the test showed that the man had died more than sixteen hundred years ago. This was no Red Christian of modern-day Denmark. This was a man of the Iron Age.

Hitchcock at his best could not do better than the mysteries written over the centuries by the peatlands of the world. In Canada, it is the muskeg. In the British Isles, moors. In the United States, we call them bogs and fens. All are only different words for places that have one key trait in common. They are places where the process of decay can't quite keep pace with the process of growth; places of slow and standing water where oxygen levels are too low to effectively break down all the remains of what has died. When something dies in a peatland, be it an Iron Age man, a leaf, or a tree, it adds to the layers of partially decomposed remains that are already there. Most of it will decay, but what does not will build up incrementally in layers over time, reaching a depth of as much as thirty feet. The layers compress under their own weight to make what we call peat, and form the basis of a unique and sometimes bizarre community where the dead are as present as the living, and the roots of plants on the surface never reach mineral soil.

I walk, on this October day, down a railroad track that runs through a bog not far from Ely, Minnesota. I listen for trains, but there is only the occasional call of a raven, the tapping of a black-backed woodpecker on a dead tree, the high purr of a half dozen cedar waxwings. The sky is clear.

The peatlands of northern Minnesota are part of a complex of boreal peatlands that ring the North Pole in North America, northern Europe, and Siberia. They are here because the last glaciers left behind a landscape of shallow depressions and flatlands where water cannot readily drain away. Their growth is nurtured by a cool, continental climate with ample rain and summer temperatures low enough that there is not an excessive loss of surface water from evaporation. Peatland nirvana.

My compass spins in circles, confused by the steel of the railroad tracks until I work my way down the embankment to enter the bog. The landscape before me is blanketed by sphagnum moss and meadows of sedges the color of old burlap with their grasslike leaves bent over into loops. Black spruce trees that look more dead than alive lift their skeletal arms as if beckoning the sky. A rain of needles falls from scattered clusters of tamarack trees turned gold in these shortening days of autumn.

I smile as I remember the words that Colonel William Byrd III wrote in his journal in 1736 about traveling through a peatland. "Never," he said, "was Rum, that cordial of Life, found more necessary than in this Dirty Place." With my first step into the bog, my right leg sinks to midcalf. Fair enough. With my second step, the water comes up, over, and inside my boot. The third step is more complicated. It seems that my right foot is fond of where it is. Would like to stay for a while. Somewhere in the neighborhood of sixteen hundred years. I begin to understand why there hasn't been much in the way of tourism in the peatlands, and why they are considered by some to be the region's last true wilderness.

I extricate my foot. Within a short distance I'm beyond the deep water that rings the bog like a moat, and moving gracelessly across the hummocks of green and burgundy sphagnum. It's a little like walking on a trampoline; people have been known to get seasick walking on a bog. The earth gives way beneath my weight. Behind me, I hear the gurgle of water as it rushes back in where my footsteps had displaced it. This place will not miss me when I'm gone.

Peatland communities offer a lean environment for a plant to make a living. As I walk, I know that I tread on nutrients held hostage in the peat—nutrients that the living plants cannot use unless they are set free by the peatland's reluctant process of decay.

The raised surface of the bog and its dense peat base further isolate the plants from the flow of runoff and groundwater. These waters carry dissolved mineral ions that would provide nourishment for plant growth and a buffering of the acids formed within the sphagnum moss. Without them, a bog is left poor in nutrients and high in acids. It is this that distinguishes the two peatland communities termed bogs and fens. A fen, by definition, lies within the path of mineral-laden waters. The surface of a fen is not raised, so it is accessible to runoff. Groundwater that wells to the surface may also travel through the low-lying plants of a fen, bringing a boost of nutrients and bicarbonate or other base from surrounding sediments that buffers the acidity of its waters.

In diversity of species, a fen is considered to be much richer than a bog. Its outside source of sustenance and more alkaline waters allow it to support a greater array of plants. A bog is "fed" only by rain, snowmelt, and what ions the winds can carry from surrounding uplands. Few species of plants can survive in a bog. Even fewer can thrive. Peatlands are often a mosaic of bogs and fens, with the bogs differentiated more by what they lack than what they contain.

Part of the intrigue of a peatland community, be it bog or fen, is the pairing of life and place. As the albatross can soar above the sea for hours without even a flap of its long wings by using the aerodynamic lift created in the friction of air and waves, so too has the life in a peatland found its own way to get along with the conditions it offers.

The black spruce and tamarack I see around me in the bog are able to grow new roots from their trunks and lower branches as the water level rises around them. When high winds throw a black spruce down, upright stems can sprout along the length of the fallen trunk to form new trees. The Labrador tea and leather-leaf plants have waxy and hairy surfaces on their leaves that may help to prevent the loss of water through evaporation during the extended winters of the north when their roots are ineffective in the frozen peat. Many plants, like the bog rosemary, have small leaves and hang on to them for a long time, saving some of the energy that it takes to produce and maintain them. Others reach beyond a diet of sun and rain. A pitcher plant lies in wait for an insect, nestled in the moss with its palmful of acid cupped in blood red leaves.

Animals also find a home in the range of habitats offered by these northern bogs and fens, or weave their days between the peatlands and surrounding upland forests. Their presence in and use of the peatlands vary with the season. Sandhill cranes nest in summer in the open fens. Male black bears seek their winter dens in tamarack or black spruce stands. In early spring, the great gray owls wing the spaces between these scattered trees, hooking lemmings and red-backed voles in their talons to feed the owlets that wait in an old raven's nest back in the recesses of the bog.

Northern leopard frogs, spring peepers, wood frogs, and boreal chorus frogs ensure that the peatlands will not lack for song; their reproductive success is greater in the fens where their eggs fare better in more alkaline waters. Bobcats and coyotes stalk their prey among the moss-covered mounds, and a myriad of songbirds live and breed in the peatlands. More than a third of Minnesota's species of birds are said to be major users of peatland habitats. Many of them, like the Connecticut and palm warblers, feed on the plentiful supply of insects that bog and fen communities provide.

Where did it begin, I wonder, the notion that a land without people is uninhabited?

I lean over to pick a ripe cranberry from its trailing stem, pop it in my mouth and cringe at the bite of juice on the back of my tongue. The flavor of bog.

A few days ago, I flew in a small plane over the more than 87,000-square-acre Red Lake Peatland north of Bemidji, Minnesota. The Red Lake Peatland is among the world's most stunning examples of what is called a "patterned peatland."

Unlike the confined peatlands that dwell in small depressions, a patterned peatland forms on broad expanses of flat or gently sloping ground. In the case of the Red Lake Peatland, formed in the ancient bed of Glacial Lake Agassiz, the "slope" represents a drop of only one to five feet per mile. Groundwater that seeps up at the upland end of the peatland, along with surface waters contributed by runoff, cannot drain through the impermeable substrate. The water has nowhere to go but to creep downslope at a rate of only a few feet in a week's time, seeking its course as topography and the plants of the peatland allow.

In this setting, in a dynamic and still not fully understood interplay between hydrology, topography, water chemistry, and climate, are created the striking landscape designs for which patterned peatlands are named. They are best seen from the vantage of a bird or a cloud.

The plane passed over Upper Red Lake, the pilot pointing out the wild rice cultivated by farmers along its margins. Then the lake was behind us. What I saw ahead through the dust of that plane's window could well have been the face of another planet.

The features on the surface, though made with a wild hand, were strangely ordered. A vast plain stretched flat and tight as a drum skin out to the circle of the horizon. From it arose islands the shape of teardrops, all oriented with their rounded heads to the west and their trailing tails to the east. Between the teardrops, the surface broke into a phalanx of quavering stripes; broad bands of gold and green that could have been the plow furrows of a farmer with a little too much coffee in his veins. A lone meandering stream and occasional mirrored flash of the sun off the surface were the only visible signs of the water that I knew was moving below the plants as a great, patient river across the landscape.

The illusion was that we were witnessing a suspended parade formation that would begin again as soon as we looked away. But the truth is that it was proceeding as we watched, our sense of time too hurried to perceive its infinitesimal steps.

Science has found names, and reasons, for these patterns. The reasons, like all reasons, are theories. The teardrops are called "tree islands." They are aligned parallel to the flow of water through the peatland, and are not islands in the traditional sense of elevated land, but rather are clusters of trees with moss hummocks at their feet. The stripes are alternating pools of water called "flarks" and ridges of peat called "strings" that are dominated by sedges. They run transverse to the migration of water. Tree islands, strings, and flarks form in the channeled water tracks of the Red Lake Peatland that are classified as patterned fens. Their ordered arrangement is thought to be primarily a response of plant growth to the distribution of nutrients carried in the water as it flows. Between the water tracks, but out of reach of its nutrients, are raised bogs whose forested crests give way to lawns of sphagnum moss.

The roaring of the engine in my ears was nothing compared to the silent roar of the landscape below the plane; in the face of our labels and hard-earned understandings, still inscrutable. Remote.

I had thought that it would feel different here on the ground. Immersed in this musky smell. Standing in the light filtering through humid air. Able to set my feet in the soft moss, to reach my hand down into the peat and close my fist around the living dead in its icy, pulpy mass. Yet it is no less elusive in its very midst, reeking of time and exquisite otherness, the way a long-time lover can suddenly seem less known than is a stranger.

I wonder when the train comes.

Boundary Waters Wilderness

January

For the listener, who listens in the snow,
And, nothing himself, beholds
Nothing that is not there and the nothing that is.
(Wallace Stevens, "The Snow Man")

If it is true that we are an accumulation of place, then I am at least in part this sharp scent of a balsam fir needle crushed in a hand. Enhancing my vision is the swirling, dizzied view of the whirligig beetle I've watched skim this lake's surface in summer, riding the interface of water and sky with two pair of eyes so that it might see at the same time what is below and what is above. When I rest, it is as the tadpole rests in spring on the branches of submerged trees at the water's edge. My shoulders have learned from these white pines how to gather in the moon. In the unspoken voice of my sometime sadness is every loon I've heard cry into the spire-circled dusk of these lakes. As does the granite, I count among the lines of my face the etchings of boulder grit carried in glaciers. And whatever part of mine it is that welcomes a bright, bitter cold, that quickens at silences too profound to break—that part, I know, came from here.

"I think I saw a wolf last night." I crack a salted peanut, drop the shells into an ashtray. Peanuts and coffee are the closest thing to breakfast that Gladys, owner of this small roadside northern Minnesota bar, has to offer. Gladys nods in response. It's mostly wolves around here, she says, and where you find wolves you won't find many coyotes. She goes into a back room and comes out with a wildlife book. She opens it to a page that compares the two, sets the book in front of me, tops off my coffee. Coyotes, according to the text, are a quarter to half the size of wolves; their ears are larger relative to their body size, their noses are more pointed, and they tend to carry their tails curled downward as they run. I look up. *"It was broad across the chest. Big. Held its tail high."* Gladys nods again. Smooths the page. Says she likes to look at the book when things are slow. The pages of the book are worn soft as cloth.

Gladys grows distant, caught in her own thoughts as she turns to put the book back in its place. Left alone, I return to the night before, to the white oval of a frozen lake illuminated by refracted starlight. It is plenty bright enough to make my way. My snowshoes break the crust of snow into jagged plates, the kind of crust that will shred the shins of a moose, ring its tracks with blood. A pine-hushed arctic wind pours out of the Quetico. Beneath the snow, the lake ice makes sounds: booms, groans. Sometimes a high-pitched keening like the singing of whales. When I pause, my breath rises in a vapor cloud that stiffens my eyebrows with frost.

On satellite photographs of North America taken at night, in which all the city lights shine like constellations, this is one of the blessed black spaces. Sprawled in the darkness before me are millions of acres of wild country, from the Boundary Waters Canoe Area Wilderness on into Canada's Quetico Provincial Park. Red and white pine of the temperate forests are joined at this latitude by upland spruce, balsam fir, and jack pine in the first glimmers of the great boreal forest that will begin in earnest to the north.

Thousands of lakes lie here in the beds made for them in the Canadian Shield's bedrock by the scouring of the last ice age. The shape of each body of water is a singular record of yielding. Long, thin lakes follow fault lines of the rock and lie in parallel valleys where glaciers raked out the ancient muds of the Rove Formation from between stronger blades of diabase. Lakes of wandering shores, irregular as Rorschach blots, fill depressions where glacial ice encountered instead more uniformly resistant rock such as gabbro and granite.

Low black hills surround the pool of light that is this lake. In silhouette, a monarch white pine lifts its arms up above the other trees, both gangly and graceful, the tall child in the school photograph. I pick out the slight saddle in the horizon that hints of a stream-carved passage to another lake. In spring, summer, or fall, it is a place I would paddle toward in a canoe, where I would find a narrow trail, the portage. Along it would be lady-slipper orchids, twinflowers, blue-bead lilies. A caution would ring in my thoughts as I crossed over patches of bare bedrock with the canoe on my shoulders: *Watch for the nighthawk cryptic on her eggs, feathers mottled gray like the rock, who would not move even as your boot descended upon her.* None of it matters now. Not the turn of the paddle blade on the recovery stroke to slice without resistance through the air, not the swing of the wind toward the east that would give warning of a wave-kicking squall, or the distance kept to give a loon peace. These are all useless on this winter night, the foreign coins left over from a trip to another season.

I seek out Aldebaran in the sky overhead, the eye-star of Taurus grown big and red with old age. In the clarity of the bone-dry atmosphere, the stars seem to have drawn nearer to earth: descended, perhaps, to peer at the peerless irony of humans who

would give the name "canoe area wilderness" to a land that more than half the year is locked in ice.

This border lakes landscape is closer to the North Pole than to the equator. Its climate is for the most part untempered by the Great Lakes, and zero is a height that winter temperatures often observe only at a distance. But more than cold, the region would better be thought of as *lean* compared to lands at more southerly latitudes. Begun with little till atop bedrock, the soils have had less time to develop since the departure of glacial ice, and have been slow to build under the conifer-dominated forest that came to establish here. Plants experience essentially drought conditions while surface waters are frozen from late September through April. The lakes are clear and beautiful but low in fertility. And the oblique angle of the sun's rays means a lessening of energy coming into the natural system. From these finite accounts must life be drawn.

Accordingly, the more cold tolerant and less energy demanding conifers are favored over deciduous trees. Many species of wildlife enter a period of dormancy for part of the year. Those that can—including 80 percent of the summer population of breeding birds—migrate during the months when there is a decline in availability of their traditional foods, such as fish or insects.

What hearts, I wonder, can the snowy owl hear beating on this mid-January night? It would be quiet compared to the cacophony of the growing season. There would be those of moose, lynx, woodpeckers, beaver, and otter, of shrews and voles in their snow tunnels, and fish making slow turns beneath the lake ice. There might even be, I realize, some hearts newly emerged into the world.

I smile to think how good are the odds that I share the passing minute with a black bear somewhere nearby in her den giving birth to cubs. She will be curled into a sandy bank overlooking a cedar swamp, or snugged up against a fan of roots in a hollow she has dug at the base of a downed tree. The calluses on the pads of her feet will have begun to wear off during her dormancy, revealing smooth new skin beneath. She will be smaller, and will likely have one or two rather than the three to five cubs of the sows fattened on the blackberries and acorns of richer habitats to the south. But the mature virgin forests that comprise half of the wilderness area will have provided enough dogwood, beetles, blueberries, and wild sarsaparilla to sustain her, just as they have provided the boreal owl with a nesting cavity in an old aspen, and the pine marten with woody debris on the forest floor where it may rest under the snow and hunt its rodent prey. These animals are among the living, breathing products of what some would call an unproductive forest.

I have stood still too long. The cold presses down like a weight. Not prepared to spend the night, I flip first one, then the other snowshoe around until I am headed back toward where I began. The trail of shallow craters made by my snowshoes has begun to drift in with the fine snow that skates across the lake's wind-packed surface.

The time would have been somewhere past midnight when I made my way out. The wolf emerged from a curtain of trees, crossing over a narrow road that dead-ends into the wilderness. It looked back—once—over its shoulder to the place where I stood. Then it slipped again into forest. Its tracks already felt cold to the touch when I knelt beside them only seconds after the wolf had left them. Maybe they had never been warm.

I leave a stack of quarters on the pine planks of the table for Gladys. Downing the last swallow of coffee, I step outside into a morning just about bright enough to shatter an eye. A raven atop a spruce lets loose with a string of quorks, trills, yells, knocks, pops, and bells. I nod in agreement with whatever it meant to say.

ECO-REGION VIII

Aspen Parkland: Brush Prairie, Aspen Groves, Beach Ridges

Lay of the Land

Eco-region VIII was once covered by Glacial Lake Agassiz and is dominated by such lake-related features as lake-bed plains, low dune and swale topography, gravel beach ridges, and rocky glacial till plains reworked by water. Aside from the beach ridges, much of the region is virtually level. The Roseau River and isolated streams meander northward toward Hudson Bay.

Native Communities

On this northern reach of the plains-forest transition zone, woody plants such as trembling aspen, bur oak, balsam fir, and willow shrubs begin to infiltrate the tallgrass prairie in a savanna-like community known as the aspen parkland. Poorly drained lowlands of Glacial Lake Agassiz are associated with a mosaic of wetland communities, including sedge meadows, calcareous fens, and wet prairie, grading in the east to extensive peatlands.

Animations

• Mounds of burrowing animals are sites of renewal in the parkland. Ground squirrels, pocket gophers, coyotes, and badgers all expose patches of bare soil. Birds such as sharptail grouse are drawn to the mounds for dust baths, leaving behind seeds— often of snowberry or other berry-producing shrubs—in their droppings. The shrubs then establish and grow, providing the protection needed for wind-blown aspen seeds to take root.

• As on the adjacent prairies, large boulders in the parkland are thought to have been used as scratching posts by bison, particularly in spring when they would have been shedding their winter coats. Although bison were extirpated from the region by the late 1800s, some of these boulders are said to still hold a shine from the polishing they received.

• Sponsors of the 1862 Homestead Act regarded land as worthless until it was "improved," that is, changed from its natural state. According to the law, individuals could receive 160 acres by living on the land for five years and making "improvements" such as clearing or plowing. Additional land could be obtained by a *tree claim,* whereby the homesteader was required to plant 675 trees for every 20-acre plot. The bias toward trees still lingers, as does the term "improved" for disturbed land, at the expense of the many native wildlife species dependent for habitat on open lands and the mosaic of open country and patchy forest.

• Minnesota's aspen parkland represents the southern corner of a larger aspen parkland landscape that extends across extensive areas of Manitoba and Saskatchewan. Its character changes somewhat in Canada, where the tallgrass prairie understory is gradually replaced by mixed-grass prairie and fescue grassland.

Border Justifications

The aspen parkland's western border is defined by the occurrence in Eco-region I of tallgrass prairie associated with the clay plain of Glacial Lake Agassiz. The border with Eco-region VII marks a transition eastward to poorly drained peatlands. To the south, the border with Eco-region III is characterized by relatively greater relief and a warming climate.

Caribou Wildlife Management Area

August

There is a place where the sidewalk ends
and before the street begins . . .
(Shel Silverstein, "Where the Sidewalk Ends")

I have a waning moon in the rearview mirror and a bank of dark clouds to my left, rising like a mountain range from the Minnesota–North Dakota border. The landscape collapses quietly as I head north. There is a gradual easing of the hills, as if someone's god were pulling the ends of the horizon line taut under the moonlight. In that one long hour that is all the hours between midnight and dawn, a whitetail buck comes up like an apparition in the headlights. He throws a crazed look in my direction and spins away. His hooves kick up a spray of gravel that machine-guns the fender. I replay the scene in my mind for the next stretch of miles, seduced by the wonder of a bad thing that could have happened and didn't.

Morning is truckstop coffee in a paper cup to go. A bad thing that could have happened and did. Canada's red maple leaves are mingled in among the license plates. Above is a sky big enough to make you feel airborne with your feet still on the ground. I look through my tapes of music to try to match it, and choose

Carreras, Domingo, and Pavarotti in Rome. On the high notes, they come close.

Maybe thirty-five miles from the Canadian border, I turn and head due east to cross the Kittson County line. The landscape is level enough that the caterpillars crawling across the road stand out as topographic relief. This is the smooth bed of Glacial Lake Agassiz, whose waters left its lake-bed plains about nine thousand years ago. Now it is part of that midcontinental stage upon which the Great Plains of the west begin their slow fade into the forests of the east. It is here that a combination of climate and other factors begins to tip the balance from grasslands to trees. Where the transition occurs along this northern latitude of Minnesota, there exists a distinctive community known as the aspen parkland.

The name was given by the first Europeans, who admired the open, parklike beauty of the landscape with its clusters of pale aspen trees and willow shrubs set in meadows of prairie grasses and sedges. But long before the Europeans these lands were known to the Cree, the Sioux, the Ojibwe, and the Assiniboine people as their hunting grounds and home. The people of these first nations would welcome the wildfires that burned here in the dry seasons, and would set their own if the wildfires didn't come. They knew that it was only in the aftermath of fire that fresh browse would grow anew to tempt the bison and wapiti.

Regular low-intensity fires are still the key to the integrity of the parkland, allowing the prairie to hold its own in a continuous tango with the forest. A fire knocks back woody plants that would otherwise take over the clearings. It doesn't eliminate trees, but rather keeps them in check. The aspens, balsam poplar, and willows are all able to sucker up again from underground rhizomes following a fire. Prairie grasses are thought to benefit from spring fires that occur before they have begun their season's aboveground growth; with litter cleared away, they are warmed by the sun on the blackened ground, and will answer a fire with more blooming flowers and greater seed production. The phrase "natural disturbance" doesn't do it justice. A fire in the aspen parkland is a *revival* in the best, most glorious, rafter-shaking sense of the word.

As I head from west to east, I look for hints of the transition happening in between the plowed fields and pastures. At first it is clearly prairie country. Except for the small fringes of green ash, basswood, and bur oak known as gallery forests that grow in fire-protected areas along the banks of the occasional stream, the only trees are those that people have planted in windbreak rows and around their farmhouses. When I start to see clusters of aspen trees on the rare idle piece of ground, I know that I've arrived.

I wind my way toward a remote wildlife management area south of the ghost town of Caribou (reported to have had an "estimated population of ten" before its grocery store closed). The roads turn to gravel. Then to miles of sandy two-track with grasses mounded high between the tire ruts. Then just to grasses.

The car's engine ticks as it cools. I strike off away from the car, which looks odd here, its technology so much cruder and less sophisticated than that of the surrounding landscape. At least twenty natural communities make up the parkland ecosystem, laid out in a matrix that is 50 to 75 percent open: primarily prairie and wet meadow. Woody plants—mainly isolated aspen groves, shrub thickets, and an occasional oak savanna—are an important but relatively sparse component of the system. The resulting structure, or *physiognomy,* is a key reason for the parkland's particular assembly of both forest and plains wildlife, who make use of different aspects of the region according to their changing needs through the seasons. Trees and shrubs offer cover throughout the year, and in winter their buds make an important addition to the food supply. The open country offers seeds to autumn migrants, as well as the long view that many plains animals need to reduce the chances of successful ambush by predators. Each successive horizontal layer—from underground, to low grasses and forbs, to high grasses, shrubs, and tree canopy —is host to its own suite of creatures for nest sites and feeding.

I take a deep breath and feel the immensity of the landscape around me. No white-noise drone of traffic. No rumble of tractors or trains. No powerlines. Kittson County is said to have 116,000 acres of aspen parkland in roughly a dozen big blocks that are mosaics of land owned by private individuals, DNR Wildlife Management Areas, Scientific and Natural Areas, a state park, and land purchased by The Nature Conservancy. About half is in public lands, half in private. It is not all protected. Nor can it all be said to be pristine, for even wild lands reflect the motives of their managers. It is nevertheless of a wholeness rare in these times, filling mile after lonely mile in the northwest reaches of Minnesota and on up into Canada.

The tracks of a moose meander through a sedge meadow that, though surely wet in spring and early summer, is now dry underfoot on this mid-August afternoon. The trail of hooves leads me through a maze of willow shrubs that reach higher than my head. The tracks are old and I do not expect them to lead me to their source. It is enough to walk where the moose walked.

Aspen parkland is in some ways considered to be atypical habitat for moose, who are generally more closely associated with areas of greater cover where their poor sight is not such a liability. They are drawn to the food they find in the wetlands here, and to the good browse of willows, aspen, saskatoon. And, no doubt, to the elbow room.

I pause. The leaves in the near aspen groves flutter on their slim petioles, rippling in the light like water over the pebbles in a shallow stream. They are old and young at the same time; old below the ground and young above. A core sample of a single tree might show fifteen annual rings. The tangle of roots below the ground, from which the tree sprang following the last fire, may have been growing for thousands of years. Answering the question of how old the aspens are is as complicated as asking how old *we* are; it is a question not so much of where a life begins, but of where one ends.

A gentian shouts purple at my feet. Ahead, behind, above, below are the thousand private worlds of the parkland. The world of the sharptail grouse feasting on winter willow buds and taking dustbaths on badger mounds. The perfumed world of a rare white orchid, the western prairie fringed orchid, visited in the night by the sphinx moth who will come for nectar and will fly away with orchid pollen on its eyes. The world of the yellow rail calling from the wet meadow with a sound like the striking of one rock on another. The world of a moose calf concealed in prairie grass, legs curled beneath it, asleep. And imbedded within each of these worlds are all the worlds that came before, that teach of migration and of adaptation to drought and cold, that learned to spin with prairie fire and glacial ice and things that we, even in the best of our dreams, have not even begun to imagine.

A wind comes up, rushing with surprising force out of the east. For the first time, I notice curtains of rain falling not far away.

Listen.

Calls of sandhill cranes. They emerge through the rain on wings that span six feet of sky. Two adults with red crowns, two young. They land in a clearing and continue to call, heads perched on slender necks snaking forward and back, the calls trumpeting, percussive. The adults have painted themselves with soil. They have preened it into their gray feathers until they were stained red-brown, as if it were part of them. Or they were part of it.

A SELECTED LIST OF SITES

Baxter's Hollow, Hemlock Draw/Baraboo Hills
Sauk County, Wisconsin
Owned and managed by The Nature Conservancy, Wisconsin
Chapter

*Beaches Lake Wildlife Management Area/Aspen Parkland
   Preserve*
Kittson County, Minnesota
Owned and managed by the Minnesota Department of Natural
Resources, assisted with property acquisition by The Nature
Conservancy, Minnesota Chapter

Blue Spring Oak Opening State Natural Area/Kettle Moraine
Jefferson County, Wisconsin
Owned and managed by the Wisconsin Department of Natural
Resources

Bluestem Prairie Preserve
Clay County, Minnesota
Owned by The Nature Conservancy, Minnesota Chapter, with a portion leased to the Minnesota Department of Natural Resources as a Scientific and Natural Area; cooperatively managed by both agencies

Boundary Waters Canoe Area Wilderness
St. Louis, Lake, and Cook Counties, Minnesota
Owned and managed by the Superior National Forest, U.S. Department of Agriculture

Caribou Wildlife Management Area
Kittson County, Minnesota
Owned and managed by the Minnesota Department of Natural Resources

Erie Marsh Preserve
Monroe County, Michigan
Owned and managed by The Nature Conservancy, Michigan Chapter

Hole in the Mountain Prairie Preserve
Lincoln County, Minnesota
Owned and managed by The Nature Conservancy, Minnesota Chapter

Horicon Marsh National Wildlife Refuge
Dodge and Fond du Lac Counties, Wisconsin
Owned and managed by the U.S. Fish and Wildlife Service

Kettle Moraine State Forest, Southern Unit
Jefferson, Walworth, and Waukesha Counties
Owned and managed by the Wisconsin Department of Natural Resources

Lost Lake Peatlands Scientific and Natural Area
St. Louis County, Minnesota
Owned and managed by the Minnesota Department of Natural
Resources, assisted with property acquisition by The Nature
Conservancy, Minnesota Chapter

Lulu Lake Preserve
Walworth and Waukesha Counties, Wisconsin
Owned and managed by the Wisconsin Department of Natural
Resources, assisted with property acquisition by The Nature
Conservancy, Wisconsin Chapter

Maxton Plains Preserve
Chippewa County, Michigan
Owned and managed by The Nature Conservancy, Michigan
Chapter

Mink River Estuary
Door County, Wisconsin
Owned and managed by The Nature Conservancy, Wisconsin
Chapter

Nerstrand Big Woods State Park
Rice County, Minnesota
Owned and managed by the Minnesota Department of Natural
Resources State Parks Division, assisted with property acquisi-
tion by The Nature Conservancy, Minnesota Chapter

Nordhouse Dunes Wilderness Area
Mason County, Michigan
Owned and managed by the U.S. Forest Service, Manistee
National Forest, assisted with property acquisition by The
Nature Conservancy, Michigan Chapter

Pointe Mouillee State Game Area
Monroe and Wayne Counties, Michigan
Owned and managed by the Michigan Department of Natural
Resources

Porcupine Mountains Wilderness State Park
Ontonagon and Gogebic Counties, Michigan
Owned and managed by the Michigan Department of Natural
Resources

St. Clair Flats Wildlife Area
St. Clair County, Michigan
Owned and managed by the Michigan Department of Natural
Resources

Sand Lake Peatlands Scientific and Natural Area
Lake County, Minnesota
Owned and managed by the Minnesota Department of Natural
Resources

Sleeping Bear Dunes National Lakeshore
Leelanau County, Michigan
Owned and managed by the National Park Service, U.S.
Department of the Interior

Warren Woods Nature Study Area
Berrien County, Michigan
Owned and managed by the Michigan Department of Natural
Resources

SOURCES

General

Aaseng, Norman, et al. 1993. *Minnesota's Native Vegetation: A Key to Natural Communities.* St. Paul: Minnesota Department of Natural Resources, Natural Heritage Program, Version 1.5 Biology Report 20.

Albert, Dennis A. 1994. *Regional Landscape Ecosystems of Michigan, Minnesota and Wisconsin: A Working Map and Classification.* Fourth Revision. Minneapolis: United States Forestry Service.

Albert, Dennis A., Shirley R. Denton, and Burton V. Barnes. 1986. *Regional Landscape Ecosystems of Michigan.* Ann Arbor: University of Michigan, School of Natural Resources.

Bakeless, John. 1961. *America as Seen by Its First Explorers.* New York: Dover.

Eichenlaub, V. L. 1979. *Weather and Climate of the Great Lakes Region.* Notre Dame, Ind.: University of Notre Dame Press.

Evers, David. 1992. *A Guide to Michigan's Endangered Wildlife.* Ann Arbor: University of Michigan Press.

Hargrave, Bryan. 1994. *The Upper Levels of an Ecological Classification*

System for Minnesota. St. Paul: Minnesota Department of Natural
  Resources.
McNab, W. H., and P. Avers. 1994. *Ecological Subregions of the United
  States: Section Descriptions.* Washington, D.C.: United States
  Department of Agriculture.
Terres, John. 1980. *Audubon Society Encyclopedia of North American Birds.*
  New York: Alfred A. Knopf.
Wisconsin Department of Natural Resources, Bureau of Endangered
  Resources. 1986. *Wisconsin Natural Areas and Natural Divisions,* PUB-
  LER-034, map.

Eco-region I

Boorstin, Daniel. 1993. *The Discoverers.* New York: Random House.
Brophy, John A. 1989. "Timetable of the Major Geological Events Which
  Shaped the Bluestem Prairie Area." Western Preserve Office of
  Minnesota Chapter of The Nature Conservancy and Moorhead State
  Science Center.
Cather, Willa. 1913. *O Pioneers!* Boston: Houghton Mifflin.
Durand, Paul. 1994. *Where the Waters Gather and the Rivers Meet.* Prior
  Lake, Minnesota.
Hazard, Evan. 1982. *The Mammals of Minnesota.* Minneapolis: University
  of Minnesota Press for the James Ford Bell Museum of Natural
  History.
Nature Conservancy. 1988. *Minnesota Chapter Preserve Guide.*
  Minneapolis: Minnesota Chapter.
Nature Conservancy. 1983-84. *Element Occurrences, Hole in the Mountain
  Preserve.* Minneapolis: Minnesota Chapter.
Ojakangas, Richard, and Charles Matsch. 1982. *Minnesota's Geology.*
  Minneapolis: University of Minnesota Press.
Schwartz, George, and George Thiel. 1963. *Minnesota's Rocks and Waters.*
  Minneapolis: University of Minnesota Press.
Searle, R. Newell. 1983. *Gardens of the Desert.* Minnesota State Park
  Heritage Series, Number 5. Minnesota Parks Foundation.
Svedarsky, Dan, and Ed Weiland. 1985. *The Prairie Chicken in Minnesota.*
  Crookston: University of Minnesota, Department of Natural
  Resources.
Wedin, David. 1992. "Grasslands: A Common Challenge." *Restoration
  and Management Notes,* 10:2 (Winter): 137-43.
Wendt, Keith. 1990. "What Is a Prairie?" *Minnesota Volunteer.* July-
  August: 22-29.

Personal Communications

Dr. Allen Ashworth, North Dakota State University, Fargo.

Terry Boerboom, Minnesota Geological Survey, United States Geological Survey, St. Paul.

Robert Dana, Minnesota Department of Natural Resources, Natural Heritage Program, St. Paul.

Paul Durand, historian/author, Prior Lake, Minnesota.

Dr. Lee Frelich, University of Minnesota, Department of Forest Resources, St. Paul.

Jerry Selby, Director of Science and Stewardship, Iowa Field Office, The Nature Conservancy.

Dr. John Toepfer, Fish and Wildlife Management Program, Little Hoop Community College, Fort Totten, North Dakota.

Brian Winter, The Nature Conservancy, Minnesota Chapter, Glyndon.

Eco-region II

Clark, Fred, Becky Isenring, and Michael Mossman. 1993. "Baraboo Hills Inventory Final Report: A Report to the Nature Conservancy." Madison: Wisconsin Chapter of The Nature Conservancy.

Davenport, Don. 1989. *A Traveler's Guide to Wisconsin State Parks.* Madison: Wisconsin Department of Natural Resources.

Eggers, Steve, and Donald Reed. 1987. *Wetland Plants and Plant Communities of Minnesota and Wisconsin.* St. Paul: U.S. Army Corps of Engineers.

Grimm, Eric. 1991. "An Ecological and Paleoecological Study of the Vegetation in the Big Woods of Minnesota." Ph.D. thesis, St. Paul, University of Minnesota Forestry Library.

Huntington, George. 1985. *Robber and Hero.* 1895. Reprint. St. Paul: Minnesota Historical Society.

Johnson, Gaylord. 1926. *Nature's Program.* New York: Nelson Doubleday.

Kane, Lucile M., June D. Holmquist, and Carolyn Gilman, eds. 1978. *The Northern Expeditions of Stephen H. Long: The Journals of 1817 and 1823 and Related Documents.* St. Paul: Minnesota Historical Society Press.

Lange, Kenneth. 1989. *Ancient Rocks and Vanished Glaciers: A Natural History of Devil's Lake State Park, Wisconsin.* Madison: Wisconsin Department of Natural Resources.

McLeod, Donald. 1846. *History of Wikonsan.* Buffalo Steele's Press.

Minnesota Department of Natural Resources, Natural Heritage Program. n.d. *Minnesota Dwarf Trout Lily.* Biological Report 18. St. Paul.

Mossman, Michael, and Kenneth Lange. 1982. *Breeding Birds of the Baraboo Hills, Wisconsin: Their History, Distribution and Ecology.* Madison: Wisconsin Department of Natural Resources and Wisconsin Society for Ornithology.

Partners in Flight. n.d. *Will We Lose Our Songbirds?* Washington, D.C.: National Fish and Wildlife Foundation.

Plowden, C. Chicheley, ed. 1972. *A Manual of Plant Names.* 3rd ed. London: George Allen & Unwin.

Reed, Donald M. 1985. "Composition and Distribution of Calcareous Fens in Relation to Environmental Conditions in Southeastern Wisconsin." M.S. thesis, University of Wisconsin, Milwaukee.

Sleutto, Bjorn. 1994. "Ice Age Survivors." *Earth,* 3, no. 5 (September): 34-39.

Taylor, J. Wolfred, ed. 1986. *The Kettle Moraine State Forest Turns Gold: A 50 Year Celebration of the Great Glacier.* Madison: Wisconsin Natural Resources Magazine.

Terborgh, John. 1989. *Where Have All the Birds Gone?* Princeton, N.J.: Princeton University Press.

Wisconsin Chapter of The Nature Conservancy. 1988. *The Places We Save.* Preserve Guide. Madison.

Personal Communications

Larry Bauer, Trout Lily Preserve, The Nature Conservancy, Minnesota Chapter, Faribault.

Kim Chapman, The Nature Conservancy, Minnesota Chapter, Minneapolis.

Dr. Ed Cushing, University of Minnesota, St. Paul.

Dr. John E. Dallman, University of Wisconsin Zoological Museum, Madison.

Bob Dott, University of Wisconsin, Madison.

Hannah Dunevitz, Minnesota Department of Natural Resources, Natural Heritage Program, St. Paul.

Tim Ehlinger, University of Wisconsin, Milwaukee.

Tom Holden, Maritime Museum, Army Corps of Engineers, Duluth, Minnesota.

Mary Jean Huston, The Nature Conservancy, Wisconsin Chapter, Baraboo.

Harold Kruse, naturalist/Baraboo Hills resident, Loganville, Wisconsin.

Ron Kurowski, U.S. Forest Service, Kettle Moraine State Forest, Eagle, Wisconsin.

Ken Lange, Devil's Lake State Park, Baraboo, Wisconsin.

Wayne Ostlie, The Nature Conservancy, Great Plains Program, Minneapolis, Minnesota.

Don Reed, Southeastern Wisconsin Regional Planning Committee, Waukesha.

Peg Robertson, U.S. Forest Service, North Central Forest Experiment Station, St. Paul, Minnesota.

Gary Werner, Ice Age Park and Trail Foundation, Madison, Wisconsin.

Eco-region III

Behler, John, and Wayne King. 1979. *Audubon Society Field Guide to North American Reptiles and Amphibians.* New York: Alfred A. Knopf.

Bolsenga, Stanley, and Charles Herdendorf, eds. 1993. *Lake Erie and Lake St. Clair Handbook.* Detroit, Mich.: Wayne State University Press.

Cain, Stanley A. 1935. "Studies on Virgin Hardwood Forest: III. Warren Woods, a Beech-Maple Climax Forest in Berrien County, Michigan." *Ecology,* 16: 500.

Curtis, J. T. 1959. *Vegetation of Wisconsin.* Madison: University of Wisconsin Press.

Daniel, Glenda, and Jerry Sullivan. 1981. *The North Woods of Michigan, Wisconsin and Southern Ontario: A Sierra Club Naturalist's Guide.* San Francisco: Sierra Club Books.

Donnelly, Gerard. 1986. "Forest Composition as Determined by Canopy Gap Dynamics: A Beech-Maple Forest in Michigan." Ph.D. dissertation. Department of Botany and Plant Pathology, Michigan State University, East Lansing.

Donnelly, Gerard, and Peter Murphy. 1987. "Warren Woods as Forest Primeval: A Comparison of Forest Composition with Presettlement Beech-Maple Forests of Berrien County, Michigan." *Michigan Botanist,* 26: 17-26.

Griffiths, Ronald, et al. 1991. "Distribution and Dispersal of the Zebra Mussel (Dreissena polymorpha) in the Great Lakes Region." *Canadian Journal of Fisheries and Aquatic Science,* 48: 1381-88.

Harlow, William. 1957. *Trees of the Eastern and Central United States.* New York: Dover.

Herendorf, Charles, et al. 1986. "The Ecology of the Lake St. Clair Wetlands: A Community Profile." Report 85 (7.7) September.

Washington, D.C.: National Wetlands Research Center, U.S. Fish and Wildlife Service, U.S. Department of the Interior.

Nalepa, Thomas, and Jeffrey Gauvin. 1988. "Distribution, Abundance, and Biomass of Freshwater Mussels (Bivalva: Unionidae) in Lake St. Clair." *Journal of Great Lakes Research,* 14: 411-19.

Nature Conservancy. 1994. *The Conservation of Biological Diversity in the Great Lakes Ecosystem: Issues and Opportunities.* The Nature Conservancy Great Lakes Program, Chicago, Illinois.

Nature Conservancy. n.d. "St. Clair Delta Site Profile." Great Lakes Heritage Data System, Great Lakes Office, Chicago, Illinois.

Warren Woods Site Committee, Michigan Natural Areas Council. 1964. *Warren Woods Site Report.* Bloomfield Hills, Mich.: Cranbrook Institute of Science.

Personal Communications

Rex Ainslie, Michigan Department of Natural Resources, Pointe Mouillee State Game Area, Monroe and Wayne Counties.

Pat Comer, Michigan Natural Features Inventory, East Lansing.

Dr. Gerald Donnelly, Morton Arboretum, Chicago, Illinois.

Dave Ewert, The Nature Conservancy, Michigan Chapter, East Lansing.

Dr. Lee Frelich, University of Minnesota, Department of Forest Resources, St. Paul.

Rex Hubbard, Warren Dunes State Park, Sawyer, Michigan.

Dr. Eugene Jaworski, Eastern Michigan University, Ypsilanti.

Ernie Kafcas, Michigan Department of Natural Resources, St. Clair County.

Dr. Donald Schloesser, National Biological Survey, Ann Arbor, Michigan.

David Weaver, Michigan Department of Natural Resources, Lansing.

Eco-region IV

Ashworth, William. 1987. *The Late, Great Lakes.* Detroit, Mich.: Wayne State University Press.

Barnes, Burton, Corinna Theiss, and Xiaoming Zou. 1992. "Patterns of Kirtland's Warbler Occurrence in Relation to the Landscape Structure of its Summer Habitat in Lower Michigan." *Landscape Ecology,* 6, no. 4: 221-31.

Cowles, Henry C. 1899. *Ecological Relations of the Vegetation on the Sand Dunes of Lake Michigan.* Chicago: University of Chicago Press.

Dorr, J. A., Jr., and D. F. Eschman. 1963. *Geology of Michigan.* Ann Arbor: University of Michigan Press.

Holman, Alan. 1990. "The Riddle of the Whales." *Michigan Natural Resources Magazine,* 59, no. 5 (September-October): 30-37.

Huron-Manistee National Forests. *Kirtland's Warbler Fact Sheet.* Mio, Mich.

Nature Conservancy Great Lakes Office. 1993. "An Analysis of Systems and Threats in the Great Lakes Basin. " Chicago.

Pielou, E. C. 1988. *The World of Northern Evergreens.* Ithaca, N.Y., and London: Cornell University Press.

Roethele, Jon. 1985. "Dunes." *Michigan Natural Resources Magazine,* 54, no. 4 (July-August): 20-36.

Thompson, Paul. 1967. "Vegetation and Common Plants of Sleeping Bear." Cranbrook Institute of Science, Bulletin 52.

Walkinshaw, Lawrence. 1987. *Kirtland's Warbler.* Bloomfield Hills, Mich.: Cranbrook Press.

Weeks, George. 1988. *Sleeping Bear: Its Lore, Legends and First People.* Glen Arbor/Ann Arbor: Cottage Book Shop of Glen Arbor and the Historical Society of Michigan.

Personal Communications

Burt Barnes, University of Michigan, School of Natural Resources, Ann Arbor.

Peggy Burman, Sleeping Bear Dunes National Lakeshore, Empire, Michigan.

Pat Comer, Michigan Natural Features Inventory, Michigan Department of Natural Resources, Lansing.

Sue Crispin, Great Lakes Office, The Nature Conservancy, Chicago, Illinois.

Dr. C. R. Harington, Ottawa, Ontario.

Rex Hubbard, Warren Dunes State Park, Sawyer, Michigan.

Dr. Allen Holman, Michigan State Museum, East Lansing.

Doug Munson, U.S. Forest Service, Huron National Forest, Mio, Michigan.

Tom Weise, Michigan Natural Features Inventory, Michigan Department of Natural Resources, Lansing.

Eco-region V

Cashatt, Everett D., and Timothy Vogt. 1992. "The Wisconsin 1991 Status Survey for the Hine's Emerald Dragonfly (*Samatochlora heneana Williamson*)." Report submitted to the U.S. Fish and Wildlife Service, January.

Catling, P. M., et al. 1975. "Alvar Vegetation in Southern Ontario." *Ontario Field Biologist,* 29, no. 2: 1-25.

Chapman, Kim. 1984. "Alvar: There's More to It Than a Strange Name." *Michigan Natural Area News,* no. 12 (January).

Corbet, Phillip. 1963. *A Biology of Dragonflies.* Chicago: Quadrangle Books.

Herendeen, Patrick, and Stephen N. Stephenson. 1986. "Short-term Drought Effects on the Alvar Communities of Drummond Island, Michigan." *Michigan Botanist,* 25: 16-27.

Keough, Janet. 1986. "The Mink River—A Freshwater Estuary." *Wisconsin Academy of Arts and Sciences, Arts and Letters,* 74: 1-11.

Martin, Lawrence. 1965. *The Physical Geography of Wisconsin.* Madison: University of Wisconsin Press.

Nature Conservancy. "Selected Site Profiles, Great Lakes Database." Nature Conservancy Great Lakes Initiative. Chicago, Illinois.

Nature Conservancy. 1993. *Michigan's Last Great Place Announced: The Northern Lake Huron Shoreline.* Michigan Chapter, East Lansing.

Nature Conservancy. *Strategic Plan, Northern Lake Huron Bioreserve.* Michigan Chapter, East Lansing.

Palmquist, John, ed. *Wisconsin's Door Peninsula: A Natural History.* Appleton, Wis.: Perin Press.

Stephenson, Stephen N. 1983. "Maxton Plains, Prairie Refugia of Drummond Island, Chippewa County, Michigan." *Proceedings of the Eighth North American Prairie Conference.* East Lansing: Department of Botany and Plant Pathology, Michigan State University.

U.S. Department of the Interior Fish and Wildlife Service. 1993. "Endangered and Threatened Wildlife and Plants: Proposed Rule to List the Hine's Dragonfly as Endangered." *Federal Register,* 58, no. 190 (October).

Personal Communications

Terry Boerboom, Minnesota Geological Survey, St. Paul.

Kent Gilges, The Nature Conservancy, Michigan Chapter, Upper Peninsula Project Office, Cedarville.

Mike Grimm, The Nature Conservancy, Wisconsin Chapter, Algoma.

Pat Hudson, National Biological Survey, Great Lakes Science Center, Ann Arbor, Michigan.

William Smith, Natural Heritage Inventory Program, Wisconsin Department of Natural Resources, Madison.

Dr. Steve Stephenson, Michigan State University, East Lansing.

Kim Wright, The Nature Conservancy, Wisconsin Chapter, Madison.

Eco-region VI

Brueton, Diana. 1991. *Many Moons.* New York: Prentice Hall.

Doolittle, T., and J. Van Stappen. 1990. *1990 Migratory Bird Survey.* Apostle Islands National Lakeshore. Bayfield, Wis.

Farrand, William R., and Christopher W. Drexler. 1985. "Late Wisconsinian and Holocene History of the Lake Superior Basin." In *Quarternary Evolution of the Great Lakes,* ed. P. F. Karrow and P. E. Calkin. Geological Association of Canada Special Paper 30. St. John's, Newfoundland.

Gill, Sam. 1992. *Dictionary of Native American Mythology.* Santa Barbara, Calif.: ABC-CLIO.

King, P. B. 1965. "Tectonics in Quarternary Time in Middle North America." In *The Quarternary of the United States,* ed. D. G. Frey and H. E. Wright. Princeton, N.J.: Princeton University Press.

Landis, Scott. 1992. "Seventh-Generation Forestry." *Harrowsmith Country Life,* November-December.

Matthiessen, Peter. 1994. *The Wind Birds.* Shelburne, Vt.: Chapters.

Meeker, James E. 1993. "The Ecology of 'Wild' Wild-Rice *Zizania palustris var palustris* in the Kakagon Sloughs, A Riverine Wetland on Lake Superior." Ph. D. dissertation, University of Wisconsin, Madison.

Michigan Nature Association. 1989. *Walking Paths in the Keweenaw.* Avoca, Mich.

Nature Conservancy. *Site Conservation Plan, Horseshoe Harbor Preserve.* Michigan Chapter, East Lansing.

Nepstad, Daniel Curtis. 1988. "Wind Related Control of Tree Form and Density on Lake Superior Ridge Crests." M.S. thesis. Department of Botany and Plant Pathology, Michigan State University, East Lansing.

Pecore, Marshal. 1992. "Menominee Sustained-Yield Management: A Successful Land Ethic in Practice." *Journal of Forestry,* 90 (7): 12-16.

Rafferty, Michael, and Robert Sprague. 1993. *Porcupine Mountains Companion.* White Pine, Mich.: Nequaket Natural History Associates.

Schwab, David. "Great Lakes Storm Surge and Seiche." NOAA Great Lakes Environmental Research Library, Ann Arbor, Mich.

Spotts, Richard. 1994. *Preliminary Analysis of Conservation Issues in the Kakagon/Bad River Sloughs Watershed.* Wisconsin Chapter of The Nature Conservancy, Madison.

Personal Communications

Jim Carstens, Michigan Technical University, Houghton.

Dr. James Clark, Calvin College, Grand Rapids, Michigan.

Pat Collins, Lake Superior Bi-national Program, Minnesota Department of Natural Resources, Cloquet.

Dr. Lee Frelich, University of Minnesota, Department of Forest Resources, St. Paul.

Dr. Charles Kerfoot, National Biological Service, Ashland, Wisconsin.

Curt Larsen, U.S. Geological Survey, Reston, Virginia.

Jim Meeker, Northland Community College, Ashland, Wisconsin.

Scott Miller, Menominee Tribal Enterprises, Keshena, Wisconsin.

Lani Poynter, Delaware Mine, Copper Harbor, Michigan.

Joe Dan Rose and Ed Leoso, Bad River Natural Resources Department, Odanah, Wisconsin.

Dave Schwab, Great Lakes Environmental Research Laboratory, Ann Arbor, Michigan.

Richard Spotts, Kakagon Watershed Project, The Nature Conservancy, Wisconsin Chapter.

Dr. Paul Sundeen, Michigan Geological Survey, Lansing.

Julie Van Stappen, Apostle Islands National Seashore, Bayfield, Wisconsin.

Eco-region VII

Berlin, N. L. 1980. "The Unique Flora of the Susie Island Archipelago." The Nature Conservancy, Minnesota Chapter, Minneapolis.

Breining, Greg. 1992. "Rising from the Bogs." *Nature Conservancy Magazine,* 42, no. 4: 24-29.

Burt, William. 1957. *Mammals of the Great Lakes Region.* Ann Arbor: University of Michigan Press.

Buskirk, Steven. 1993. "Conserving Circumboreal Forests for Martens and Fishers." *Conservation Biology,* 6, no. 3 (September).

Coles, Bryony. 1989. *People of the Wetlands: Bogs, Bodies and Lake Dwellers.* New York: Thames Hudson.

Fish and Wildlife Service, U.S. Department of the Interior, Environmental Protection Agency. 1987. *The Ecology of Patterned Boreal Peatlands of Northern Minnesota: A Community Profile*. Biological Report 85, June.

Frelich, Lee. 1993. "Fire Policy in the BWCAW: Discussion of Some Ecological Issues." *BWCA Wilderness News* (August): 6-7

Glaser, Paul, and H. E. Wright Jr. 1983. *Postglacial Peatlands of the Lake Agassiz Plain, Northern Minnesota,* ed. Lee Clayton and J. T. Teller. Geological Association of Canada Special Paper 26.

Glob, P. V. 1969. *The Bog People*. London: Faber and Faber.

Green, John. 1977. *Environmental Geology of the North Shore*. Minnesota Geological Survey (U.S.G.S.) and Coastal Zone Management Program, State Planning Agency.

Johnson, Charles W. 1985. *Bogs of the Northeast*. Hanover, N.H.: University Press of New England.

Johnson, Nancy. 1984. "Susie Island, Minnesota's Arctic Outpost." *Minnesota Monthly,* June.

Larsen, James A. 1982. *Ecology of the Northern Lowland Bogs and Conifer Forests*. New York: Academic Press.

Lindquist, Edward, and Lynn Rogers. 1992. *Supercanopy White Pine and Wildlife,* ed. Stine and Baughman. St. Paul: University of Minnesota, Department of Forest Resources, College of Natural Resources.

Luoma, Jon, and John Shaw. 1981. "Requiem for a Lonely Wilderness." *Audubon,* 83 (September): 112-27.

Lydecker, Ryck. 1982. *The Edge of the Arrowhead*. Minnesota Sea Grant Extension Program, University of Minnesota.

Minnesota Department of Natural Resources. 1984. *Recommendations for the Protection of Ecologically Significant Peatlands*. November. St. Paul.

Nero, Robert. 1988. "Denizens of the Northern Forests." *Birder's World,* 50, no. 292 (September-October): 20-25.

Ojakangas, Richard, and Charles Matsch. 1982. *Minnesota's Geology.* Minneapolis: University of Minnesota Press.

Phillip, Michael J. 1981. "Peatlands." *Sierra Magazine,* September-October.

Sanger, Jon. 1987. "Remote Islands in Lake Superior Shelter Unique Plants." *Minnesota Volunteer,* May-June.

Sansome, Constance Jefferson. 1983. *Minnesota Underfoot*. Stillwater, Minn.: Voyageur.

Wright, H. E., Barbara Coffin, and Norman Aaseng, eds. 1992. *Patterned Peatlands of Minnesota.* Minneapolis: University of Minnesota Press.

Personal Communications

Chel Anderson, Tofte, Minnesota.
Kim Chapman, The Nature Conservancy, Minnesota Chapter, Minneapolis.
Dr. Magaret Davis, Department of Ecology and Behavioral Biology, University of Minnesota, Minneapolis.
Dr. John Dickey, University of Minnesota, Minneapolis.
Nick Duff, Encampment Forest Association, Two Harbors, Minnesota.
Marshall Helmberger, Raven Tours, Tower, Minnesota.
Dr. David Mech, National Biological Survey, St. Paul, Minnesota.
Dr. Robert Nero, Manitoba Ministry of Natural Resources, Winnipeg, Manitoba.
Karen Noyce, U.S. Department of Fish and Wildlife, Grand Rapids, Minnesota.
Dr. Lynn Rogers, U.S. Forest Service (retired), Ely, Minnesota.
Dr. Constance Sansome, Carleton College, Northfield, Minnesota.
Mary Shedd, U.S. Forest Service, Kawishiwi District, Minnesota.
Steve Wilson, Minnesota Department of Natural Resources, Scientific and Natural Area Program, Eveleth, Minnesota.

Eco-region VIII

Bird, Ralph. 1961. *Ecology of the Aspen Parkland of Western Canada.* Winnipeg: Research Station, Canada Department of Agriculture.
Cuthrell, David. 1994. *Insects Associated with the Western Prairie Fringed Orchid, Platanthera praeclara Sheviak & Bowels, in the Sheyenne National Grassland.* Fargo: Department of Entomology, North Dakota State University.
Kittson County Historical Society. *Kittson: A County in Motion.* Lake Bronson, Minn.
Sather, Nancy. 1993. "Minnesota's Other Northland: The Aspen Parkland." (Nature Conservancy) *Minnesota Chapter Newsletter,* Winter.

Personal Communications

Robert Dana, Minnesota Department of Natural Resources, Natural Heritage Program, St. Paul.

George Davis, Minnesota Department of Natural Resources,
   Northwestern Region, Karlstad.
Gordon Dietzman, International Crane Foundation, Baraboo, Wisconsin.
Don Faber-Langendoen, Midwest Office, The Nature Conservancy,
   Minneapolis, Minnesota.

Laurie Allmann has been a regularly featured essayist and commentator for Minnesota Public Radio's "Voices from the Heartland" series, and selections of her poetry have been published in the *Floating Fish Quarterly Review*. She has adapted her writing to the stage in performances at the Playwright Center and the Southern Theater in Minneapolis. Her professional background includes work as a wilderness canoe guide and naturalist. Allmann was born in Jordan, Minnesota, and now writes from her home in the St. Croix River Valley. *Far from Tame* is her first book.

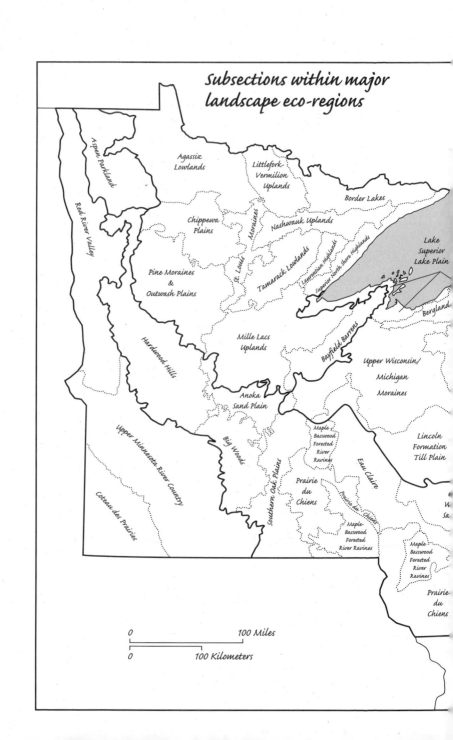

Subsections within major landscape eco-regions

Aspen Parkland

Red River Valley

Agassiz Lowlands

Littlefork-Vermilion Uplands

Border Lakes

Chippewa Plains

St. Louis Moraines

Nashwauk Uplands

Laurentian Highlands

Superior North Shore Highlands

Lake Superior Lake Plain

Pine Moraines & Outwash Plains

Tamarack Lowlands

Bergland

Hardwood Hills

Mille Lacs Uplands

Bayfield Barrens

Upper Wisconsin/ Michigan Moraines

Anoka Sand Plain

Maple Basswood Forested River Ravines

Lincoln Formation Till Plain

Upper Minnesota River Country

Big Woods

Southern Oak Plains

Prairie du Chiens

Eau Claire

Prairie du Chiens

Coteau des Prairies

Maple Basswood Forested River Ravines

Maple Basswood Forested River Ravines

Prairie du Chiens

| 0 | 100 Miles |
| 0 | 100 Kilometers |